srivatsan

Path Level Traffic Grooming in WDM Optical Networks

srivatsan

Path Level Traffic Grooming in WDM Optical Networks

Network design and Performance Analysis of Path Level Traffic Grooming Mechanisms in WDM Optical Networks

VDM Verlag Dr. Müller

Impressum/Imprint (nur für Deutschland/ only for Germany)
Bibliografische Information der Deutschen Nationalbibliothek: Die Deutsche Nationalbibliothek
verzeichnet diese Publikation in der Deutschen Nationalbibliografie; detaillierte bibliografische
Daten sind im Internet über http://dnb.d-nb.de abrufbar.

Coverbild: www.purestockx.com

Verlag: VDM Verlag Dr. Müller Aktiengesellschaft & Co. KG
Dudweiler Landstr. 125 a, 66123 Saarbrücken, Deutschland
Telefon +49 681 9100-698, Telefax +49 681 9100-988, Email: info@vdm-verlag.de
Zugl.: Ames, Iowa State University, Diss., 2007

Herstellung in Deutschland:
Schaltungsdienst Lange o.H.G., Zehrensdorfer Str. 11, D-12277 Berlin
Books on Demand GmbH, Gutenbergring 53, D-22848 Norderstedt
Reha GmbH, Dudweiler Landstr. 99, D- 66123 Saarbrücken
ISBN: 978-3-639-06724-8

Imprint (only for USA, GB)
Bibliographic information published by the Deutsche Nationalbibliothek: The Deutsche
Nationalbibliothek lists this publication in the Deutsche Nationalbibliografie; detailed
bibliographic data are available in the Internet at http://dnb.d-nb.de.

Cover image: www.purestockx.com

Publisher:
VDM Verlag Dr. Müller Aktiengesellschaft & Co. KG
Dudweiler Landstr. 125 a, 66123 Saarbrücken, Germany
Phone +49 681 9100-698, Fax +49 681 9100-988, Email: info@vdm-verlag.de

Produced in USA and UK by:
Lightning Source Inc., 1246 Heil Quaker Blvd., La Vergne, TN 37086, USA
Lightning Source UK Ltd., Chapter House, Pitfield, Kiln Farm, Milton Keynes, MK11 3LW, GB
BookSurge, 7290 B. Investment Drive, North Charleston, SC 29418, USA
ISBN: 978-3-639-06724-8

1

TABLE OF CONTENTS

4

DEDICATION

I would like to dedicate this book to my father Mr. V. Balasubramanian and my mother Ms. R. Padma without whose support I would not have been able to complete this work.

6

ACKNOWLEDGEMENTS

It has been a long and winding road. But, it has been education all along. I would like to take this opportunity to express my thanks to those who helped me with various aspects of conducting research and the writing of this thesis. First and foremost, I thank Prof. Arun K Somani for his guidance, and support throughout this research and the writing of this thesis. I would like to acknowledge Prof. Ahmed E Kamal, who is an excellent teacher and who guided me a lot during the initial stages of my graduate career. I would also like to thank my committee members for their efforts: Prof. Robert Weber, Prof. Fernandez Baca and Dr. Lu Ruan. I would additionally like to thank the graduate school and the department for giving me an opportunity for pursuing my graduate studies here. This research work is funded in part by NSF grant CNS 0434872 and 0626741, Information Infrastructure Institute at Iowa State University (ICUBE), and Jerry. R. Junkins endowment

I would like to thank all my friends who made my life exciting during my graduate days: Satyadev Nandakumar, Bhuvaneswari Ramkumar, Mahadevan Gomathisankaran, Harisudhakar Vepadharmalingam, Vikram Sriram, Samarth Shetty, Girish Lingappa, Veerendra Allada, Muthuprasanna Muthusrinivas, Srinivas Neiginhal, Venkatesh Selvaraj, Kanaga Karuppiah, Muthukumar Kadavasal, Abhijit Rao, Aswin Natarajan, Ganesh Subramanian, Jing Fang, Yana Ong, Ravi Cherukuri, Kamna Jain, Kavitha Balasubramanian, Kalyani Siddarth, Kishore Ramachandran, Vyjayanthi Prasad, Sumitha Alex, Neeraj Koul, Rahul Marathe, Kousik Ganesan, Souvik Ray, Varun Sekhri, Rohit Gupta, Durga Kocherlakota, Sankalpites, Oaklanders, the Pikeys, the chemi gang, the coffee room gang, the sankalp gang and the MTV gang. A special mention here of Navin Parthasarathy, Karthikeyan Balasubramanian and Kripa Karthikeyan.

CHAPTER 1. IP over Optical Networks - Introduction

The telecommunications industry has been witnessing an exponential growth of network traffic in the past few years. The aggregate bandwidth requirement of the Internet is expected to be well over 5000 petabits/day by 2007. While voice traffic growth has been slow for many decades, there has been a surge in the growth of data traffic and data is expected to be over 75 % of the total network traffic seen in the Internet. With the continuing proliferation of bandwidth-intensive multimedia applications and widespread availability of broadband access technologies, this paradigm shift in capacity demands is having a profound impact on today's network design and deployment.

Telecommunication networks can be roughly organized into a three-tiered hierarchy: access, metro and long haul [49]. The access networks provide the subscriber interface to the communication network. It hosts a broad range of protocols/technologies and supports a wide variety of application devices. On the other end of the hierarchy is the long haul, which provides large tributary connectivity between regional and metro domains. There has been an unanimous agreement among backbone service providers that Dense Wavelength Division Multiplexing (DWDM) offers the best cost-capacity trade-off and hence is the technology of choice for the long haul. Interfacing the access with the long haul is the metro. The metro segment provides high speed media and application devices required to interconnect the access networks to the core. The emerging trends in traffic have significantly altered the domains bordering the metro and have made service providers seriously rethink the current technologies that are in place.

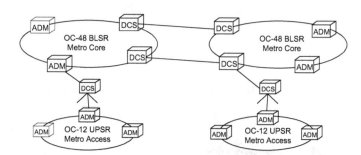

Figure 1.1   Current SONET based metro networks: OC-12 UPSR metro access rings and OC-48 BLSR metro core rings

1.1   Traditional Metro Architectures

Currently, metro networks are based upon SONET/SDH ring architectures and are organized into a two-level hierarchy: metro edge (or metro access) and metro core. The metro edge refers to the space between subscriber access and central office location. Metro edge rings span about 10 to 40 kms, operate at OC-3/STM-1 or OC-12/STM-4 rates and employ Add Drop Multiplexers (ADMs) that connect to digital loop carrier setups, enterprise networks, telephone public branch exchanges etc. Most edge traffic is usually outbound from the local ring and hence exhibit strongly hubbed traffic patterns [49] making edge networks well suited for UPSR architectures.

The metro core refers to the rings that interconnect major central office hub locations and that feed into long haul networks. Metro core rings span about 40 to 80 kms, operate at OC-48/STM-16 or OC-192/STM-64 rates and perform a higher level of aggregation than the corresponding edge rings. The traffic demands in metro core are much more meshed and improved bandwidth efficiency is obtained through BLSR architectures. Digital cross-connects that can switch in both space and time are used to interconnect rings and to provide fine granular bandwidth management. The traditional ring architectures shown in Figure 1.1 performed well when the dominant traffic was voice. However, there have been some emerging trends (discussed in Section 1.2) in the metro space that bring to the forefront the inherent deficiencies in existing architectures.

1.2 Emerging Networking Trends

Growing demands

The tremendous growth in internet traffic volumes is fueled by content-rich applications like packetized voice, internet gaming, video on demand, and streaming multimedia. New services that are offered include interconnecting and consolidating data centers and transparent extension of the LAN across the MAN. There is a trend towards supporting SAN architecture, real-time transactions backup, high-speed disaster recovery, grid computing and the more futuristic optical virtual private networks. Concurrently, there remains a very healthy demand for legacy voice and leased-line services, arising from a huge, entrenched base. It is important to note that the bursty nature of IP data traffic requires that network design be different from the conventional telephony design.

Advancing access technologies

Many technologies are emerging in the access domain including cable, DSL, high speed wireless, wavelength leasing and wavelength on demand. Improved access technologies make possible wide spread use of bandwidth-intensive applications which in turn create the need for more efficient access networks thereby entering a positive regenerative cycle. Thus, there is need for a scalable, robust and easy to manage network architecture that can support multiple access technologies and provide intelligent handling of broadband user data flows.

Asymmetry in data flows

Asymmetry in data traffic has been observed on both national and international links [77]. The cause of such asymmetry is attributed to large server farms sending out huge data packets upon receipt of small requests. The consequence of observed asymmetry is that it brings to question the efficiency of SONET based transport networks, which provisions bi-directional circuits that are severely underutilized in one of the directions and heavily congested in the other direction.

Importance of transparency

Transparency is one of the key requirements of a future-proof network in the sense that networks need not be redesigned from scratch due to changes in protocols and technologies. An all-optical network is transparent to bit rates, modulation formats and protocols and can support higher bandwidths without making massive equipment changes known as 'forklift upgrades' that require complete overhaul of existing infrastructures. This enables metro operators to scale their networks to meet customer requirements and enhance their service velocity. Elimination of electronics to do signal regeneration in the intermediate nodes lowers costs and power consumption. It also simplifies operations, since there is no need to manage disparate network elements. It offers support for legacy services and gives operators the ability to bundle services with different optical quality-of-service and service-level agreements. This feature allows the service providers to tailor service offerings to meet the needs of specific customers.

Migrating to mesh

Traditional telecommunication networks were configured as rings since they guarantee recovery and lead to predictable restoration paths thereby simplifying management. Fiber usage can be low in ring solutions because of the requirement for protection fibers on each ring. A mesh physical topology is more efficient when the demand pattern is also meshed. Besides, network designs rarely resemble rings since fibers can only be routed only along rights-of-way. Building rings on top of meshed fibers results in logical overlay which are harder to design and maintain. Mesh networks allow a topology similar to fiber routing. All links in a ring need to have the same capacity, and hence during upgradation, all the links may have to be upgraded simultaneously. In the case of mesh networks, incremental upgrades are possible. Similarly, the number of wavelengths need to be the same on all links in a ring whereas it is not so constrained in mesh networks.

Also, the benefits in flexibility and efficiency of mesh networks are potentially great. Protection and restoration can be based on shared paths, thereby requiring fewer fiber pairs for the same amount of traffic and can lead to efficient wavelength utilization. However, mesh networks will require a high degree of control and management plane intelligence to perform the functions of protection and bandwidth management, including fiber and wavelength switching.

Increasing need for reconfigurability

Conventional networks are circuit switched and are interconnected by leased lines with long holding times. However, there is an increasing need for reconfigurability in optical networks that allow bandwidth creation in real time between end users to accommodate dynamically changing traffic demands. The routers and switches should acquire the ability to set up circuits of wavelength or sub-wavelength granularity across optical backbones within seconds. Such provisioning will allow customers to buy high bandwidth for short-term use, such as a high-definition video transmission that a television network might need.

1.3 New Metro Solutions

In light of above trends, SONET based metro networks are facing serious limitations [49, 48]. We describe some of the challenges faced by the conventional networks and then critically assess some of the promising solutions that are currently available.

1.3.1 Metro Requirements

With growing demands, the capacity exhaust problem gains significance. Capacity upgrade in SONET is possible either through deploying new rings or through increasing TDM rates. The former requires new fiber routes while the latter necessitates equipment upgrades on all ring nodes both of which are expensive and time consuming. In SONET, each transport path has a fixed bandwidth defined over a rigid rate hierarchy. This leads to large bandwidth inefficient mappings while supporting a multitude of diverse client data applications. Besides, the burstiness in traffic cannot be handled well since re-provisioning requires careful capacity planning and takes long time. The network supports only constant bit rates and is not transparent, thereby providing very little room for service differentiation. So, a need for a transparent, cost-effective architecture that responds to dynamic traffic and allow for service differentiation while supporting legacy services is being increasingly realized.

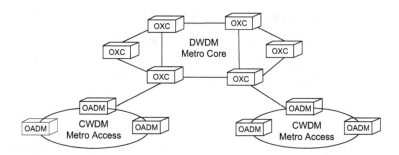

Figure 1.2   Next generation metro networks : CWDM based hubbed metro access rings and DWDM based metro core mesh networks

1.3.2   Metro Core Solutions

The requirements for metro core is different from that of metro edge. In the metro core, the emphasis is on scalable bandwidth provisioning. With maturing optical technologies, ring- or mesh-based wavelength routed DWDM networks is an ideal fit here since it offers rapid provisioning, service transparency and low network costs (since they are amortized over a large user base). The trend towards mesh networking in the metro core is reported in [90].

1.3.3   Metro Edge Solutions

In the metro edge, the focus is on protocol heterogeneity, heavily sub-wavelength traffic and a price-sensitive limited user base. Hence the metro edge is seeing more diverse possibilities, ranging from improved SONET/SDH and Ethernet offerings to optics-based propositions. We discuss each of them in turn below.

1.3.3.1   Next Generation SONET

Recently, a new standard called NGS [25] was defined to support data over fiber services while still retaining its original protection and performance monitoring features. The techniques that enable efficient data mapping include the Generic Framing procedure (GFP), Link Capacity Adjustment Scheme (LCAS) and virtual concatenation (VCAT) mechanisms. VCAT allows for concatenation of several payloads to provide flexible bandwidth and to minimize mismatch in data and port rates. GFP provides a simple framing technique [16, 115, 99] to multiplex multiple client protocols and LCAS specifies a control mechanism to dynamically adjust the number of tributaries assigned to a connection. Collectively, these features are the building blocks of the new data-aware characteristics of NGS transport networks.

Despite the above enhancements, NGS is still an approach that attempts to bridge the packet and circuit switching paradigms, both of which differ fundamentally in their philosophies. NGS systems process the signals electronically at some of the intermediate nodes thereby precluding transparency, reducing scalability and leading to increased equipment costs. Besides, NGS also has some framing requirements like STRATUM timing [113] and pointer processing which can become expensive at high data rates like 40 Gbps.

1.3.3.2   Next Generation Ethernet

The features that are exclusive to SONET is its efficient support for survivability and performance monitoring. Ethernet services, on the other hand, are easily upgradeable and has the advantages of familiarity, simplicity and low cost. While Ethernet does not offer TDM-level guarantees for bandwidth and delay, SONET does not offer efficient data mappings. NGE is a ring based fault tolerant data transport solution that combines statistical multiplexing along with a fairness based medium access scheme called Resilient Packet Rings (RPR) [35].

However, there are some problems associated with the packet scheduling and rate adaptation approach followed by RPR. The scheduling stream gives priority to transit traffic over local traffic and hence delay seen by a node is dependent on upstream traffic patterns. In addition, if the bandwidth requirement of a newly arriving traffic is lowest among the contending traffic flow, this causes all the upstream nodes to throttle their rate to this lowest rate, creating large oscillations in bandwidth allocation. Such a reactive approach in the presence of bursty traffic may result in large settling times for the oscillations. In general, packet rings have been designed based on enterprise requirements and consequently there is less support for TDM traffic. Since RPR terminates traffic on some intermediate nodes like NGS, their capacity scalability and cost-effectiveness is also questionable.

1.3.3.3   Wavelength Division Multiplexing

WDM is the sole technology that can support TDM, data, SAN, cable video etc. independent of bit rates and protocol formats. The transparency feature of WDM gives it an
edge over all other technologies. Although the other alternate solutions presented above may
delay the deployment of WDM systems, it appears to be the most scalable solution in the long
term. Since, the traffic volumes in the metro edge may not be excessively heavy as in the core,
coarse WDM (CWDM) which does not place stringent requirements on optical equipment can
be deployed, thereby leading to significant cost savings [48]. CWDM will allow operators to
expand service offerings, support legacy services and prepare for future traffic growth.

1.3.4   Next Generation Metro Networks

WDM technology provides transparency, survivability, and scalable bandwidth provisioning, which will be of great use in the metro arena. We feel that the next generation metro
networks are likely to be CWDM rings with Optical Add Drop Multiplexers (OADMs) in the
metro access along with DWDM mesh networks with Optical Crossconnects (OXC) in the
metro core as shown in Figure 1.2.

An OADM allows a signal to be added or dropped from the CWDM channels in the
network. It also provides a cost-effective means for handling pass through traffic. An OXC
can switch the optical signal coming in on a wavelength of an input fiber link to the same
wavelength on an output fiber link. A signal that enters a node without being sourced or
terminated at that node can be optically passed through. This saves on the number of Optical
Line Terminals (OLTs) since now a circuit does not require an OLT on intermediate nodes
that are not involved in a communication. An OLT is responsible for terminating an optical
signal, multiplexing/demultiplexing it and adapting the signal to the required specifications.

Given the large existing base of SONET rings, the first step in the migration process may
be to upgrade the metro core to DWDM rings after which metro access may be upgraded to
CWDM rings. Finally, the metro core topologies may become more meshed and is likely to be
more beneficial in the future.

1.4   IP - The Client Layer

Internet service has traditionally been carried over a SONET based telecommunication infrastructure that was primarily designed to carry circuit switched services over a time-division multiplexing (TDM) transport. These networks have evolved over time to become a complex multi-layered protocol stack that is extremely difficult to manage, provision, and update. The IP infrastructure was initially constructed using three layers of networking equipment. The development of DWDM technology added another layer to the network equipment architecture- a fourth layer allowing service providers to multiplex several wavelengths onto a single fiber potentially giving them limitless ability to expand. The factors [20] that originally motivated the multilayered stack shown in Figure 1.3 are:

1. Some unique functionality was implemented in each layer and so their presence was indispensable. For instance, IP layer allowed protocol interoperability while ATM switches provided access, integration, signaling, quality of service and traffic management. SONET offered multiplexing, transport, advanced operations and monitoring facilities while WDM tapped the tremendous bandwidth in the fiber.

2. The speed of the network elements at the higher levels were low and so grooming needed to be done at multiple lower levels before finally multiplexing them onto the fiber. For example, Speed: DWDM > SONET > ATM > IP . Multiple IP flows were combined into ATM VCs which in turn were multiplexed onto SONET circuits before finally grouping them into one fiber using DWDM multiplexers.

3. Consumers from different sectors had different preferences and all preferences had to be supported. The end system users preferred IP, while the enterprise backbone was primarily ATM based. The long haul carriers were predominantly using SONET while the core deployed WDM.

In the cases of traditional telecommunications carriers, the majority of equipment was originally designed for voice traffic, with IP capabilities supported as an afterthought. So, the existing multilayered architecture simply cannot scale to support the future requirements of

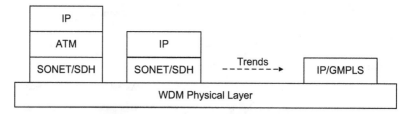

Figure 1.3   Current trends in the control plane

the Internet [61]. The multilayered architecture is frought with other problems as outlined below.

1. The multi-layered architecture complicates the timely flow of resource and topological information. The functionalities like routing and survivability are implemented on all the layers. The multilayered stack with its duplicated functionalities in different layers, far from being helpful may prove detrimental in circumstances like a link failure. For instance, in the case of a failure, each layer tries to recover using its own protocol which may create confusion in the system. For instance, hold off timer mechanisms at higher layers may be required to inter-operate optical layer survivability mechanisms with SONET Automatic Protection Switching schemes and IP rerouting strategies.

2. A single failure in the lower layer leads to multiple failures in the higher layer [61]. For instance, a single fiber cut may affect 64 wavelengths that are multiplexed onto it. If each wavelength operates at 2.5 Gbps, a total of 1000 OC-3 circuits may be affected. This will in turn translate to 10^5 VCs and 10^8 flows at the higher layers. The number of alarms generated at the higher layer due to a failure at the lower layer could be potentially impossible to manage.

3. SONET rings require 50 % of the bandwidth to be reserved for protection which lead to inefficient fiber utilization.

4. SONET proved to be a good transport network for voice traffic but is not every efficient in transporting data traffic. Since SONET does not support bandwidth provisioning

through signaling, circuit provisioning is a slow process and bandwidth reservation needs to be done during circuit set up which makes it ineffective for data traffic.

The complexity of the equipment in multiple layers is both costly to manage and difficult to update. Over time, technology has matured to design ASIC-based IP routers to run at wire speeds and yet meet various quality of service requirements and achieve carrier grade availability and reliability. This eliminates the need for ATM/frame relay switches to overcome the packet-forwarding limitations of software-based routers.

The commercial availability of OC-48c/STM-16 and OC-192c/STM-64 router interfaces obviates the need for SONET multiplexers to match the speed of router interfaces to the optical transmission infrastructure. Besides, Generalized MPLS (GMPLS) allows traffic engineering and fast rerouting thereby eliminating the need for SONET automatic protection switching and ATM traffic engineering.

The result is a slim IP over optical network that consists of a multi gigabit, label switched router (LSR) that runs IP-based Generalized Multiprotocol Label Switching (GMPLS) (described in the next section) as the control plane and that directly connects to OXCs through ports with optical transmitters and receivers. The evolution toward a two-layer network architecture shown in Figure 1.3 is currently underway as IP/MPLS and DWDM emerge as the two dominant technologies in recent network installations. The two-layered network increases service velocity and is easier to manage and optimize for performance. Besides, building the control plane from standard and proven routing and signaling methods as in GMPLS, minimizes risk and time to market for the service providers and hence is likely to develop as the widely accepted solution.

1.4.1 Generalized Multiprotocol Label Switching

In a MPLS network [93], packets are forwarded through a Label Switched Path (LSP), that is explicitly set up by signaling protocols such as RSVP-TE (Resource Reservation Protocol with Traffic Engineering extensions) [3] or CR-LDP (Constraint based Routed Label Distribution Protocol) [63]. RSVP is an IP protocol for signaling resource requirements of an application to intermediate routers. CR-LDP enables control message distribution for establishing LSPs and utilizes constraint based routing.

The path begins at an ingress Label Switched Router that prepends a label to a packet and forwards it to the next router along the path called the Label Switched Router (LSR). The LSR swaps the packet's outer label for another label and forwards it to the next LSR. This continues until the packet reaches an egress LSR, where the outer label is removed and the packet is delivered to the appropriate client network.

MPLS is based on the idea of decoupling the forwarding plane from the IP header. This allows MPLS to be not just restricted to routers but be extensible to devices like OXCs that do not have the capability to parse IP headers. The MPLS control plane operates in terms of label swapping and forwarding abstraction. The MPLS forwarding plane allows for link-specific realizations of this abstraction. It is through this possibility of abstraction, Generalized MPLS (GMPLS) allows the concept of label to be extended and defined as a wavelength in the context of optical networks. It is important to note that there are a few features that are unique to optical networks and may require suitable modifications in the control plane. A simple example [9] is that the MPLS label space is large (one million per port) whereas the number of wavelengths and TDM slots is small (tens to hundreds per port). To address this issue, GMPLS introduces a concept called LSP hierarchy, that allows LSPs to be nested inside other LSPs. MPLS LSPs that have the same entry and exit points in the optical domain are aggregated into a single optical LSP. Thus, nesting of LSPs help conserve the number of wavelengths used in the optical domain.

The enhancements that are made to GMPLS to adapt it to the optical networking paradigm include:

- A new Link Management Protocol [42] designed to address issues related to link management in optical networks using optical crossconnects.

- Enhancements to routing protocols [71, 72] like OSPF and IS-IS to advertise availability of optical resources in the network.

- Enhancements to signaling protocols like RSVP-TE [3] and CR-LDP [63] to allow explicit

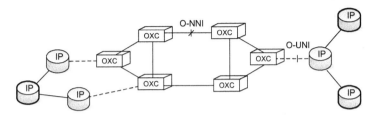

Figure 1.4 Ip Over WDM networks - IP routers connecting through an optical cloud

set up of an LSP in an optical core [14].

- Scalability enhancements such as hierarchical LSP formation and link bundling.

1.4.2 IP Over WDM - Interconnection Models

The internet model is now evolving towards wire speed IP routers connecting through an optical cloud as shown in Figure 1.4. The interaction between the client IP network and the optical network is called the Optical User Network Interface (O-UNI) and the interaction within the optical network is called the Optical Network to Network Interface (O-NNI). Based on the interaction between the IP and the optical domain [89], the services offered by the system can be categorized as Domain Service Model (DSM) or Unified Service Model (USM). In the DSM model, O-UNI is different from O-NNI whereas in the USM model, both are defined to be the same. Based on the architectural alternatives for routing information exchange between IP routers and optical switches, three different models [89] are proposed: overlay, augmented and peer model. This classification is based on the amount of control information exchanged between the two layers.

1.4.2.1 Overlay model

In the overlay model, the IP and optical networks are completely decoupled and a separate instance of a control plane runs on each network. The IP/MPLS routing protocols are completely independent of the routing and signaling protocols of the optical layer. The client can query the optical layer over the O-UNI and get information about whether a particular connection between a router pair can be set up over the optical network. The IP/MPLS router maintains information regarding the residual capacities available on the existing logical topology and the available transceiver ports and router capacities. A request may be routed over the existing virtual topology or through new lightpaths. If the router decides not to open up new lightpaths, it has to identify the virtual links over which the LSPs are to be routed. In the case, it wants to open up new lightpaths, it decides the LSR pairs between which the lightpaths are to be provisioned. The overlay model, does not exchange any information across layers and is easy to manage. But, it may lead to severe network underutilization.

1.4.2.2   Peer model

In the peer model, also known as the integrated model, the IP and optical network run a unified control plane. An interior gateway protocol like IS-IS or OSPF with suitable optical extensions may be used to exchange topology information [89]. Signaling protocols like LDP and RSVP along with their extensions can be used for setting up and tearing down LSPs and lightpaths. The LSRs have information regarding the optical network topology, the unused wavelengths and residual capacities on the logical links and transceiver ports in the network. This information is exploited to solve the LSP provisioning problem in the IP layer and the RWA problem in the optical layer in an integrated manner.

The peer model has a good potential to yield cost effective network solutions. But it is limited by the fact that LSRs and WXCs are typically managed by different vendors running proprietary protocols making the tight coupling scenario less practical. Also, it requires huge amounts of control messages to be flooded among network entities to maintain global network status.

1.4.2.3   Augmented model

The middle ground between the overlay model and the peer model is the augmented model. The augmented model requires that some amount of useful information be exchanged between the optical and electronic layer. The control information is to be small enough to prevent large flooding, and large enough to let the LSR make useful WDM aware routing decisions. An example control information type that can be exchanged across layers and that will provide good network utilization at low control overhead is investigated in [75]. The augmented model is made possible by running an optical version of Exterior BGP between the routers and the OXC over the O-UNI and between neighboring OXCs over the NNI. An optical version of Interior BGP is used between border OXCs within the same subnetwork [89].

1.5 Contributions of this work

In this chapter, we presented an introduction to technological issues related to metro optical networks and architectural issues related to IP over WDM networks. The issues discussed here set the context for the research work that we discuss in the forthcoming chapters. The goal of this dissertation is to design algorithms and protocols for cost-effective IP data transport in metropolitan optical networks. The contributions of the dissertation, organized in various chapters, is described as below:

In Chapter 2, we introduce the concept of grooming which refers to the mechanisms to pack low bandwidth requests into high speed circuits. We survey the various architectures that have been proposed in the past and how they achieve efficient network utilization through grooming. We describe the hardware and software components required to achieve such grooming capabilities and bring out the merits and demerits of each approach.

In Chapter 3, we introduce the PLATOON architecture and the various candidate crossconnect configurations that we propose for the metropolitan networks. The candidate architectures that we compare in our study are lightpath based networks and variants of light-trail based networks. We define the static trail routing and wavelength assignment problem, provide ILP formulations, resolve the complexity class of the optimization problem and propose simple polynomial time heuristics. We impose physical layer constraints and equipment capability constraints and observe the network utilization under such conditions. We also consider the dynamic connection provisioning problem. We introduce an auxiliary graph approach that can model heterogenous optical networks with different transceiver, wavelength, converter, and grooming constraints. This graph based approach is a generic method that can model all networks that allow aggregation of connections on a circuit along a linear path. This model allows for policy based grooming and also accounts for client layers with constrained electronic speeds.

In Chapter 4, we evaluate the performance of various network architectures based on the algorithms developed for both static and dynamic traffic scenarios. We obtained optimal solutions for small networks using the ILP formulations. Our algorithm for single-hop static design yields up to as high as 90 % wavelength utilization in the presence of non-splittable fractional traffic. Our simulation results for the static two-hop design suggests that with only a small number of hub nodes, high network throughput and good wavelength utilization can be achieved. Based on our study of multi-hop networks, we are able to make conclusive statements on the right type of architectures to use depending on the resource constraints in the network. We also observe that with constrained router capacities, light-trails under certain circumstances, can perform better than lightpath networks that employ electronic grooming.

In Chapter 5, we describe some of the survivability mechanisms that have been studied in the past for lightpath based networks. We provide ILP formulations for dedicated protection in light-trail networks and propose connection level shared and dedicated fault tolerance schemes for static non-splittable, subwavelength traffic. We observe, based on our simulation results, that with only a modest amount of spare capacity, full protection can be achieved with all single link failures. We develop a backup multiplexing based solution for survivable network design in the presence of dynamic traffic. We observe that survivable light-trail networks perform several orders of magnitude better than other candidate architectures that we studied.

In Chapter 6, we present a simple medium access control protocol for light-trail networks that uses fiber delay loops to buffer optical packets and avoid collisions. We use queueing theory to model the delay incurred by packets in a small network and use simulation results to model larger networks. We observe that our model is in agreement with our simulation results. We conclude that for moderate loads, our protocol works well and under certain conditions, it always works better than a protocol that was suggested earlier for the same architecture.

In Chapter 7, we present our conclusions from the research conducted in this dissertation and discuss the possibilities for future research.

CHAPTER 2.   Traffic Grooming in Optical Networks

Grooming is a terminology that captures a variety of problems in telecommunication networks that aim to optimize capacity utilization. In an abstract generic sense, it is a complex multi-commodity network flow problem that may involve one or more layers within the same system. The motivation for grooming arises because of the ability to share resources among multiple entities that need the resource. This sharing is possible and even desirable because the resources under question are expensive, individual entities need only a fraction of the resource and multiplexing allows the resource cost to be amortized over the number of users. Grooming improves resource utilization and leads to a scalable network design. Grooming is required typically in networks where the request sizes are small as compared with the circuit bandwidth.

A feature that is fundamental to grooming is the ability to switch low speed traffic streams into high speed bandwidth trunks. In the context of optical networks, it typically involves switching of traffic from one wavelength, waveband, time slot, fiber, or cable to another [12]. The general objective of grooming is to help decompose hard circuit provisioning problems into small, simpler ones and yield an increased solution space for such problems.

Given the generic nature of the problem, grooming can be provided within a layer or across layers. We call the grooming functionality that is built into the optical layer, optical grooming (o-grooming) and the grooming functionality that is available between the optical and the client layer, electronic grooming (e-grooming). In this chapter, we first review the various o-grooming techniques proposed in literature and the architectures and protocols required to achieve all-optical grooming for unicast traffic. Multicast architectures are beyond the scope of this current work and for an account of recent developments in this field, readers are referred

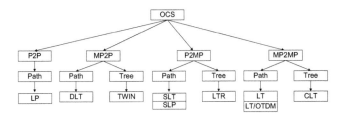

Figure 2.1   The different flavors of OCS - lightpath (LP) [146], destination
based light-trail (DLT) [17, 24], time-domain wavelength interleaved networking (TWIN) [124], source based light-trail (SLT)
[40], super lightpath (SLP) [22], light-tree (LTR) [95], super
light-tree (SLTR) [87], light-trail (LT) [52, 131], light-trail with
OTDM (LT/OTDM) [31] and clustered light-trail (CLT) [53]

to [66, 65]. Next, we survey e-grooming along various dimensions and describe some of its
basic limitations. The motivation of this chapter is to summarize some of the general trends
and techniques in this rich field of research.

2.1   Optical Grooming

The optical layer is typically equipped with reconfigurable switching fabrics and the optical
grooming technique used depends on the granularity and time scale of the switching functionality available in the optical layer. Optical switching can be done at various levels of granularity
and time scales:

1. Waveband switching

2. Wavelength switching

3. Sub-wavelength time slot level switching

4. Burst switching

5. Flow switching

6. Packet switching

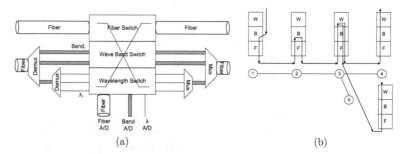

Figure 2.2 (a) The multi-granularity OXC allows hierarchical switching at the fiber, waveband and wavelength level (b) Band aggregation in an example network

Waveband level switching is a technique that is used when the optical layer switching devices are limited to operate at the coarse granularity of a set of wavelengths. If the set corresponds to all the wavelengths in the fiber, it is called fiber switching. Wavelength level switching techniques allow circuit switched sharing of a wavelength and when supplemented with some additional hardware and an overlaid control protocol (as described in later sections) can allow statistical sharing of a wavelength as well. In wavelength level switching, the optical switch is configured for the lifetime of a connection. For sub-wavelength time slot level switching, the wavelength is divided into T fixed time slots and the optical switching fabric is reconfigured for every k time slots ($k = 1..T$). For packet/flow/burst switching, the optical switching fabric is reconfigured for every packet/flow/burst, thereby ensuring that the resources are used only for the time they are required. The feature common to all the grooming strategies is switch reconfiguration at various time scales to enable efficient packing of small requests into high bandwidth pipes. There also exists a technique called code division multiplexing that uses orthogonality of codes instead of switch fabric reconfiguration to groom requests. We give a brief overview of each of these techniques in the forthcoming sections.

2.1.1 Waveband Switching

In this section, we discuss waveband switching technologies that allow for wavelengths to be bundled into bigger units called wavebands using a multi-granularity OXC (MG-OXC).

The work in [83] introduces an OXC architecture shown in Figure 2.2 (a) that allows for routing at the fiber, band and the wavelength level simultaneously. While the figure shows statically configured switch, a dynamically configured design is discussed in [83]. The MG-OXC is composed of three stages - fiber (FXC), band (BXC) and wavelength (WXC) Crossconnects. All the incoming fibers are connected to the FXC. Some of the fibers are directly switched to the output fiber ports while the rest of it is switched to the BXC fabric. At the BXC fabric, the incoming fibers are demultiplexed into its constituent bands using a fiber-to-band demultiplexer. A few of the bands are dynamically sent to the WXC fabric, where, a band to wavelength demultiplexer is used to switch at the wavelength level. This hierarchical switching reduces the port requirements and consequently leads to a scalable network design.

To design a multigranularity optical network, wavelengths need to be grouped into bands and fibers. If there are k wavelengths from a source to a destination, they can be grouped into a band since the individual wavelengths do not require preferential treatment at the intermediate OXCs. One can also aggregate traffic with different source or destination nodes but with a common subpath to form bands on this subpath. Figure 2.2(b) shows an intermediate grouping strategy suggested in [83]. Wavelengths going from node 1 to node 4 are grouped into a band along with wavelengths going from node 1 to node 5 on subpath $1 - 2 - 3$.

2.1.2 Wavelength Switching

Wavelength Switched Networks (WSNs) rely on an all-optical paradigm called Optical Circuit Switching (OCS) that is based on the notion of setting up dedicated circuits between nodes through one or more OXCs to facilitate communication requests. Many flavors of OCS have been researched in the past. Circuits are typically configured in the form of a path or a tree, and can be categorized into four main classes as shown in Figure 2.1 based on the traffic aggregation techniques used.

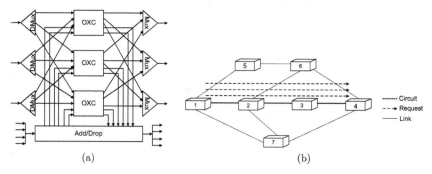

(a)   (b)

Figure 2.3 (a) The wavelength plane cross-connect supports
adding/droping of subwavelength circuits directly and
switching at wavelength granularity (b) An example LP
network

1. Point to Point (P2P): This technique allows a lightwave circuit to aggregate traffic between the convenor node and the end node of the circuit.

2. Point to Multi-point (P2MP): This technique allows a lightwave circuit to aggregate
traffic from a source to multiple destinations.

3. Multipoint to Point (MP2P): This technique allows a lightwave circuit to aggregate traffic
from multiple sources to the same destination.

4. Multipoint to Multi-point (MP2MP): This technique allows a lightwave circuit to aggregate traffic from multiple sources to multiple destinations.

We review some of the o-grooming paradigms in WSNs below and identify the aggregation
method used in each of these techniques.

2.1.2.1   Lightpath Architecture (LP)

In the optical layer, data transferred from one node to another does not get buffered in the
intermediate nodes. In the absence of wavelength converters, a single all-optical wavelength

23

continuous path is traversed from source to the destination. Such a circuit is called a lightpath
(LP). A lightpath is realized in an optical network through using optical line terminals (OLT),
optical add/drop multiplexers (OADM) (in rings) and optical crossconnects (OXC) (in mesh).
The OXC architecture shown in Figure 2.3(a) is called single hop grooming OXCs. Such OXCs,
switch at the granularity of a wavelength and has data ports that can support low speed traffic
streams directly from client network equipment [146].

The low speed streams are electrically multiplexed in the Time Division Multiplexing
(TDM) fashion. Since, the OXCs do not switch at a granularity less than that of a wavelength, all the low speed streams on wavelength channel are switched to the same destination
node. Consider the example network shown in Figure 2.3(b) that we intend to use throughout
this chapter for illustration purposes. For the rest of the chapter, the following assumptions
are valid regarding the example network unless explicitly stated otherwise. Each arriving connection at this example network is of unit size while the capacity of each circuit is assumed to
be 10 units. Let $R_{i,j}$ be the request from node i to node j. Initially, the network does not carry
any traffic and each link has one available wavelength. The requests are depicted by dotted
lines, while the links and circuits are depicted by lines of light and dark shades respectively.

Say, request $R_{1,4}$ arrives at the network. This can be carried by the circuit $C = \{1, 2, 3, 4\}$
as shown in the Figure 2.3(b). Consider two other requests between nodes 1 and 4 that arrive
later. In LP, they can be accommodated on the same circuit C. If a request arrives between
any other node pair, a separate circuit needs to be established. Thus, LP allows the circuit to
be shared by connections between the same source and destination and hence corresponds to
the P2P aggregation technique mentioned above. Since each connection traverses only one LP
circuit to reach the destination, the architecture is called a single hop lightpath architecture.
The biggest advantage of this architecture is the simplicity of the control plane and the maturity
of technologies related to hardware and software components of this architecture.

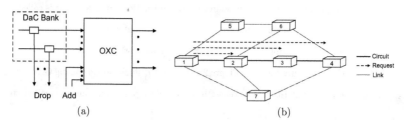

Figure 2.4 (a) The Dac switch taps signals on every incoming channel and hence is able to support dropping and extension in all-optical fashion (b) An example DaC network

2.1.2.2 Drop and Continue Architecture (DaC)

The optical crossconnect architectures are of two types: (a) terminate or continue (ToC) (b) drop and continue (DaC). In the first type, a lightwave circuit that is entering an OXC port either terminates at that node or continues to the next node and is the same as the OXC architecture for LP described in 2.1.2.1. This means that the circuit is established between the end nodes and intermediate nodes on the circuit cannot access the data on the signal. The drop and continue architecture shown in Figure 2.4 (a), however, allows for a small fraction of the power on all signals to be tapped at all intermediate nodes. Such splitting can be achieved using passive devices like optical couplers. This splitting allows the circuit originated by the head node to be shared by connections that are destined for intermediate or end nodes of the path. This is also called the P2MP aggregation strategy. A drawback of this architecture is the losses the signals may suffer due to power being tapped on all the intermediate nodes.

The architecture suggested in [40] introduces the dropping and extension concepts and is explained as follows. Consider the example shown in Figure 2.4 (b). When request $R_{1,3}$ arrives at the network, circuit $C = \{1, 2, 3\}$ is established. When the next connection $R_{1,2}$ arrives, it is served by the same circuit through the intermediate tapping feature and is referred to as *dropping* in [40]. Next, suppose, a connection $R_{1,4}$ arrives, there are two ways of accepting the call. One method is to set up a new circuit from node 1 to node 4. The other method

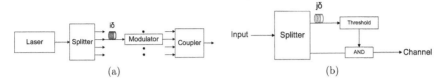

Figure 2.5   OTDM circuit for (a) transmitter and (b) receiver for bit interleaved multiplexing in the optical domain

is to dimension the existing circuit to form the new circuit $C = \{1, 2, 3, 4\}$ and this referred to as *extension* in [40]. This architecture requires burst mode receivers to lock phase with the transmitter repeatedly. For reasons to be explained in Section 2.1.2.7, we refer to this architecture as source based light-trails(SLT). The SLT architecture that we consider in our work supports only *dropping* feature and not the *extension* feature.

2.1.2.3   Optical Time Division Multiplexing Architecture (OTDM)

OTDM splits the wavelength bandwidth into a fixed number of subchannels, using a TDM scheme directly in optical domain. Bit interleaved multiplexing at an aggregate rate of 100 Gbps is possible using a mode locked laser [92] that can generate a periodic train of picosecond length pulses. This stream is split and a copy is created for each data stream to be multiplexed as shown in Figure 2.5 (a). The pulse for the ith data stream is delayed by $i\delta$ using delay lines of appropriate length. The undelayed pulse is used for framing. Each pulse is modulated externally using the local data stream. The outputs of all the modulators are combined through the star coupler to produce the desired bit interleaved optical TDM stream.

For demultiplexing, the incoming stream is split into two streams using a splitter as shown in Figure 2.5(b). To extract j^{th} stream, one of these streams is delayed by $j\delta$. The clock and framing information is obtained from the delayed stream using a threshold operation. A logical AND operation between the framing pulse stream and the non delayed stream is performed to obtain the jth stream.

The scheme proposed in [78] uses OTDM technology to form a TDM lightpath called super-lightpath (SLP). This allows the head node of the circuit to transmit data to all or a subset of the receivers that are on the superlightpath, thereby employing a point to multipoint aggregation strategy. Whenever data is desired by a local receiver, a part of the incoming optical signal is split by a coupler and tapped at the local node while the rest of the power is sent to the next node in the super-lightpath. Thus, super lightpath allows for a P2MP communication. This can lead to reduction in the number of transmitters per node and increased wavelength utilization.

In the super-lightpath approach, since there is only one source on a circuit, there is no requirement for time synchronization among nodes in the network. However, the transmitter has to operate at a speed that is equal to the aggregate speed of all the receivers in the circuit. The electronic buffering requirements for aggregating data from the clients may be excessive and expensive on the source node of the SLP. The control electronics for implementing OTDM technology is also expensive and complex.

The work in [22] propose simple algorithms to allocate user connection requests in a dynamic traffic scenario and shows that it can either lead to reduced network cost or significantly improved network utilization. The ILP formulation and heuristics for static design of such networks are studied in [78]. Based on simulation results, the authors of [78] show that the number of wavelengths required to overlay the same logical topology on the same physical topology is reduced by more that 65 % using super-lightpaths as compared with traditional lightpaths.

2.1.2.4 Light Trees Architecture(LTR)

LTR is a point to multipoint generalization of a lightpath. Through a light tree, traffic from a source is delivered to destination nodes along the tree. A 2 x 2 multicast capable wavelength routed switch is shown in Figure 2.6 (a). The information on each incident link is demultiplexed into individual wavelengths by the demultiplexer. Signals that do not need duplication are switched out to the ports where they are multiplexed to be sent out on a fiber. The signals that need to be duplicated are sent to a splitter bank. The splitter bank makes

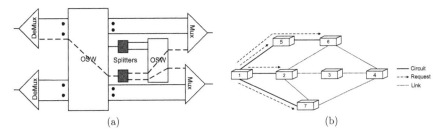

Figure 2.6   (a) The signals to be multicast are sent to the splitter banks
where the signals are duplicated and switched to appropriate
ports. The dotted lines show the path taken by a multicast
signal. (b) An example light tree network

multiple copies of the same signal and is connected to a smaller optical switch which routes

the duplicated signals to their respective output ports. This architecture [95] is similar to

the wavelength convertible switch, except that the converter banks are replaced with splitter

banks. In a light tree, the node that originates the traffic is called the root node and the

nodes that sink the traffic are called the leaf nodes. Though this architecture was originally

proposed to support all-optical multicast traffic, it can be used to carry fractional unicast

traffic as well [95]. Optical splitters leads to large amount of power losses in the signal that

needs to be compensated by using expensive optical amplifiers. An optimal MILP formulation

for power constrained design in the context of multicast wavelength routed mesh environment

is presented in [57].

Figure 2.6 (b) shows how the multicast capability can be used for grooming subwavelength

requests. Suppose, there are four low speed streams $R_{1,2}$, $R_{1,5}$, $R_{1,6}$ and $R_{1,7}$. A light tree is

established with node 1 as the root and all the requests are shared on the tree shown in Figure

2.6(b). The leaf nodes of the tree receive all the data but relay only the data intended for itself

to its client equipment while ignoring the rest. Thus, light trees, allow a circuit to be shared

by all connections which are sourced by a root node and destined for leaf nodes on a tree.

The problem of finding an optimum virtual topology is formulated as an MILP optimization

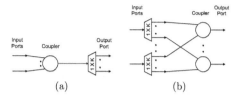

Figure 2.7   Wavelength selective crossconnect with merging (a) Full merging (b) Partial merging

problem, using principles of multicommodity flow for routing of light-trees in [95].

2.1.2.5   Super-Light Tree Architecture(SLTR)

The concept of a super-light tree was introduced in [87] and is a combination of LTR and
SLP technologies. In SLTR, a bit level OTDM scheme explained in Section 2.1.2.3 is used
and the bandwidth is split over a tree like topology using optical splitters. The tree is rooted
at a source node and reaches out to all the required destinations and carries multiple traffic
flows. The logical topology that results has increased number of logical links, thereby, reducing
average distance between nodes and alleviating congestion on logical links.

A MILP formulation and two heuristics were provided to solve the routing and wavelength
assignment problem in SLTR networks in [87]. The authors of [87] conclude that the number
of wavelengths required to lay a specified logical topology is 70 % lower than that is achievable
with traditional lightpath based networks. SLTR is however an expensive technology since it
needs optical splitters, optical amplifiers as well as OTDM electronics.

2.1.2.6   Time-Domain Wavelength Interleaved Networking Architecture (TWIN)

TWIN [124] transport network consists of Wavelength Selective Crossconnect (WSXC)
devices in the optical core and aggregation devices like IP routers at the edge (AD). Each
destination is associated with a tree that spans all the nodes in the network and is assigned
a unique wavelength. Thus, the MP2P aggregation technique along a tree like topology is

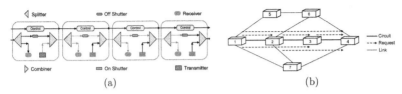

Figure 2.8 (a) A uni-directional optical light-trail circuit configured in a ring network (b) An example light-trail network

used in this architecture. When a source transmits data to a specific destination, it uses the wavelength dedicated for that destination. The WSXC fabrics are responsible for routing the wavelengths in the right direction. Data from various sources headed for the same destination is merged at several points in the network. It is important that signals from these different sources not schedule the transmission at the same time to avoid any conflicts while the signals are merged. The work in [124] proposes a distributed schedule and a centralized scheduler to support both synchronous and asynchronous traffic.

The WSXC architecture is different from that shown in Figure 2.3(a) because of the requirement of merging optical signals. An example architecture is shown in Figure 2.7 (a) where a passive coupler is used to merge all incoming signals. The 1 X K switch routes each wavelength to a distinct port. When wavelength availability is severely restricted, input signals of a wavelength may be routed to multiple output ports as shown in Figure 2.7 (b). A merging WSXC is easier to implement as compared with a normal WSXC as it does not involve complex wavelength plane switches. Besides, merging WSXCs need not be reconfigured at the packet or burst level. TWIN, however, requires each source to be equipped with fast tunable lasers and burst mode receivers to synchronize with the transmitted bit stream.

2.1.2.7 Light-trail Architecture (LT)

The LT circuit allows wavelength sharing at the optical layer among multiple source destination pairs that reside on circuit's path using a simple carrier-sensing MAC protocol discussed in [52, 4, 24, 17, 117, 69]. Several testbed implementations of variants of this architecture are reported in [116, 24, 86].

Figure 2.8(a) shows a lightwave circuit [52] established between first node (convenor node) and the fourth node (end node) on a single wavelength ring (link from node 4 to node 1 not shown). A ring topology is chosen for the sake of clarity in illustration and we extend such an architecture for arbitrary mesh topologies in Chapter 3.

Each node in the example ring consists of a splitter, a shutter and a combiner. A packet sourced from a node on this circuit traverses each of these optical components on every node en route to the end node as shown in Figure 2.8(a). At the splitter of a node, a part of the incoming signal power is sent to the local node for possible reception using a detector while the rest of the signal passes to the shutter. The shutter is a simple mirror-based optical attenuator configured to either block or let the wavelength pass through. For the first and last node on the circuit, the shutter is configured in the off/blocking position, isolating this communication circuit from the rest of the network. For intermediate nodes on the circuit, the shutter is in the on/pass-through position, letting the signal pass through. The signal, if not blocked by the shutter, travels through the combiner before exiting the node. The combiner allows the data sourced by the intermediate node to be coupled into the wavelength based on some MAC strategy. Since multiple transmitters and receivers can be present on the same circuit, light-trails employ the MP2MP aggregation strategy.

The architecture utilizes an out of band control channel, which is dropped and processed at each node to actuate the shutters. The signaling channel carries information pertaining to circuit set up, tear down and dimensioning, and can provision optical connections, ranging in duration from IP bursts to virtual circuits [52]. The shutters are not reconfigured dynamically for every packet but is done on a longer time scale thereby alleviating the switching speed requirement. This statistical sharing of a circuit leads to a better utilization of wavelength in the presence of heavily fractional traffic.

Consider the example network shown in Figure 2.8. When request $R_{1,4}$ arrives, the circuit $C = \{1, 2, 3, 4\}$ is established. This circuit is capable of supporting any subset of the set of connections $\{R_{1,2}, R_{1,3}, R_{1,4}, R_{2,3}, R_{2,4}, R_{3,4}\}$ since the capacity of the circuit is in excess of the

Architecture	Grooming Type	Grooming Capability	Medium Access Arbitration	Burst Mode Receivers	Packet Delays	Technology maturity (Hardware and Software)
LP	P2P	Low	No	No	Low	High
SLT	P2MP	Medium	No	Yes	Medium	Medium
DLT	MP2P	Medium	Yes	Yes	High	Low
LT	MPMP	High	Yes	Yes	High	Low

Figure 2.9   Comparison of variants of light-trail architectures

sum of the sizes of requests in any such subset.

Variants of Light-trails

Many variants of the light-trail architecture have been studied in literature. In all variants of the light-trail architecture, the convenor node is provisioned with a transmitter and the end is node is provisioned with a receiver. A light-trail network may be configured so that there is strictly one transmitter but possibly multiple receivers on each lightwave circuit. An instance where this P2MP aggregation technique may be used is in hubbed rings to distribute data from the hub node to the ring nodes [116, 40, 17]. We call such circuits, source based light-trails (SLT) [6] since there is only one source but multiple destinations in such a circuit. SLT is advantageous because there is no requirement for a medium access protocol. The circuit can also be configured so that there is one receiver and multiple transmitters. An instance where this multipoint to point aggregation technique may be used is in hubbed rings to collect data from ring nodes to be sent to the hub node [116, 17, 24]. We call such circuits destination based light-trails (DLT).

The characteristics of communication through light-trail circuits [5] can be explained more formally as follows. Consider a network topology as a directed graph G(V,E), with V as the vertex set and E as the edge set. Let a circuit instance be a simple path in a graph and be defined by the set $C = \{v_1, v_2, .., v_n\}$ such that $v_1, v_2, ..., v_n \in V$ and $(v_1, v_2), .. , (v_{n-1}, v_n) \in E$. Let R be the request matrix that denotes the value of the request between any node pair. A light-trail circuit can carry multiple connection requests subject to the following two basic constraints:

Containment Constraint:

This constraint specifies the type of connections supported on the circuit. If the circuit is an LT, it can support any request (v_i,v_j) if v_i, $v_j \in C$ and v_j is downstream of v_i in C. If the circuit is an SLT, it can support any request (v_1,v_j) if $v_j \in C$. If the circuit is a DLT, it can support any request (v_i,v_n) if $v_i \in C$. Using the same terminology, LP can be defined as a circuit that supports any request (v_1,v_n).

Capacity Constraint:

The sum of the traffic supported by a circuit is at most the capacity of a wavelength. If $C = 5$, $R_{v_1,v_2} = 3$, $R_{v_1,v_3} = 3$, $R_{v_2,v_3} = 2$ and the other requests are zero units, then if the circuit is a LT, it can support one of the following: $\{(v_1,v_2), (v_2,v_3)\}$, $\{(v_1,v_3), (v_2,v_3)\}$, $\{(v_1,v_2)\}$, $\{(v_1,v_3)\}$, or $\{(v_2,v_3)\}$. It cannot support, for instance, the set $\{(v_1,v_2), (v_1,v_3)\}$.

The additional constraints on the circuit may include:

Request Assignment Constraint:

A node pair's request may not be allowed to be split across multiple trails assuming that individual connections do not require more than the capacity of a wavelength. In this case, a request is assigned to exactly one circuit.

Trail Length Constraint:

In architectures like LT, the data signal incurs a power loss on every node of the path. Let trail length be defined as the number of hops in an LT. Due to power budget constraints, the maximum trail size may be limited.

While the first two constraints define light-trail circuits, the latter two constraints may depend on the services offered by the optical network and the system engineering constraints that are specific to an implementation.

The transmitter (receiver) of a node is said to be *busy* on a circuit, if the corresponding equipment is sourcing (sinking) at least one connection on the circuit. Otherwise, it is said to be *idle*. A node is said to be *active* [5] on a trail if either its transmitter or its receiver are active on the trail. It is said to be *passive* if it lies in the path of the circuit but does not have an active transmitter or receiver on the trail. A node is not *involved* in a trail if it does not lie

33

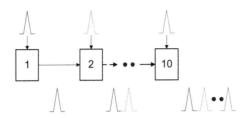

Figure 2.10   A light-trail circuit with OTDM flavor

in the path of the trail.

In SLT, scheduling of all packets is done internally on the circuit convenor node. Service differentiation is easier and round trip time delays can be avoided in SLTs. Due to distributed MAC in LTs and DLTs, the propagation delays in the feedback loop and the requirement to achieve coordination across multiple nodes, the control plane becomes more complex. The buffer requirements of LT, DLT and SLT based networks are stringent since the medium is statistically shared by multiple connections and buffers should be dimensioned to account for the contention that happens across flows. In SLT and DLT networks, the load at the convenor and end nodes are high respectively and hence the buffer requirements are high. LT circuits offer distributed aggregation and distribution. However, contention from geographically separate flows require that more buffers be allocated. The control plane for LPs is simple to manage, but the number of options available for sharing is less and hence may lead to lower network utilization. The architectures differ in the tradeoffs involved in terms of hardware requirements, performance and control complexity. A summary of the comparisons of the different variants of the light-trail architecture is presented in Figure 2.1.2.7.

Prior work in LT based networks

The RingO project [24] introduces a MP2P aggregation mechanism in a ring architecture for metro applications. The proof of concept network has nodes that consists of a tunable transmitter and a fixed receiver. To communicate with a node, data has to be sent on a wavelength dedicated for that node. Each wavelength is tapped on every node and is used for discerning activity on the wavelength. Packets from multiple sources headed for the same destination are scheduled appropriately to avoid collisions.

The authors of [20] extend the ideas in [24] to introduce the MP2P aggregation architecture in mesh networks. They provide an ILP formulation for static network design in [18] and a simple heuristic in [20] for MP2P aggregation. In [19], an analytical model for evaluating the cost of P2P and MP2P strategies in Manhattan Street Network are compared in terms of number of consumed wavelengths and transceivers under uniform static traffic. The transceiver requirements of MP2P mechanisms under static traffic scenario is discussed in [21]. The DBORN architecture proposed by the same authors in [17] proposes a dual bus architecture based on [24] for metro networks. The network is configured in the form of a ring with a hub node. The hub node employs SLT and DLT to source and sink all traffic in the ring. The fairness issues for accessing the bus are dealt with in [17].

The light-trail concept is introduced in [52]. The strategy used is an MP2MP strategy as opposed to the MP2P strategy used in [20] and [24]. A fiber level switching architecture is discussed in [51] and light-trail architecture is shown to have better network utilization and lower blocking as compared with OBS based networks. A protection and restoration scheme for light-trail ring networks is described in [55]. Five heuristics for static network design in light-trial ring networks are presented in [50]. A burstponder card implementation for ethernet grooming in light-trail WDM networks is explained in [54, 86].

2.1.2.8   Light Trails with OTDM Architecture (LT/OTDM)

The key idea behind OTDM is to have a circuit share connections from the same source to multiple destinations along a simple path. This approach has a cost disadvantage. For example, consider a network where receiver speeds are 10 Gbps but a line speed of 100 Gbps is required. In this case, each transmitter needs to be equipped with 10 modulators each operating at 10 Gbps. If less than 10 receivers along a path can utilize the circuit, some of the modulators on the source transmitter remains unused.

The idea behind light trails with OTDM flavor shown in Figure 2.10 is to have the OTDM circuit share all unidirectional connections along a simple path [31]. The transmitters along the

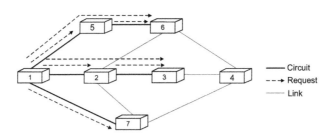

Figure 2.11   An example CLT network

path require to be synchronized for bit interleaving. The line rate depends on the number of slots per frame. With 10 Gbps per connection, a line rate of 100 Gbps can be achieved by having 10 slots per frame. With LT/OTDM, each transmitter requires only one modulator. Since multiple sources can transmit onto the same circuit, it may be easier to fill up the wavelength capacity. However, in both SLP/OTDM and LT/OTDM, the lasers need to generate pulses of width less than 10 ps for every 100 ps. In the example, ten node OTDM/LT network shown in Figure 2.10, node i transmits during time slot i to achieve an aggregate line speed of 100 Gbps.

2.1.2.9   Clustered Light Trails Architecture(CLT)

CLTs present a generic framework to which Light-trails belong and uses a MP2MP strategy along a tree like topology. The clustered light trail is a sub-tree of the network, rooted at a convenor node. An upstream node can talk to any node that is downstream to it in the tree. Consider a CLT circuit set up as shown in Figure 2.11.This circuit is capable of supporting any subset of the set of connections $\{R_{1,2}, R_{2,3}, R_{1,3}, R_{1,7}, R_{1,5}, R_{5,6}, R_{1,6}\}$ since the capacity of the circuit is in excess of the sum of the sizes of requests in any such subset.

The medium access could be either through a scheduling method or through a MAC protocol. The MAC protocol suggested in [53] gives more importance to an upstream transmitter than to a downstream transmitter. The formulation of a linear program to establish CLTs in a network with given set of traffic demands is described in [53].

36

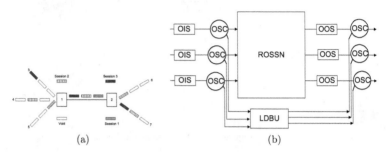

Figure 2.12 (a) An example WDM/TDM network (b) The architecture of
an optical time division space switch

2.1.3 Sub-Wavelength Switching

Sub-wavelength switching networks typically involve switch reconfiguration time scales
much shorter than what is typically found in WSNs. If a circuit switched optical layer is
used, the wavelength is divided into time slots and fixed bandwidth is allocated to each incoming connection. The bandwidth allocated to a connection can be increased on demand by
increasing the number of slots. An architecture that allows for switching at the time slot level
is described below.

2.1.3.1 Time-Wavelength-Space Routers based Networks (TWSRN)

The work in [58] introduces a circuit switched time division multiplexed wavelength routed
network in which a wavelength is composed of frames each of T time slots. Consider an example
network shown in Figure 2.12(a) that has one wavelength on every link and three time slots
per frame. Assume that each request requires one time slot and three requests $R_{3,7}$, $R_{5,7}$, $R_{4,6}$
arrive. In conventional wavelength routing, one of the requests could have been carried since
only one wavelength is available on link $(1,2)$. The switch at node 1 allows the requests on
the wavelengths from different links to be groomed onto a single wavelength on link $(1,2)$ and
the switch at node 2 allows the traffic to be bifurcated. A node with such a switch structure is
called Time-Wavelength-Space Switched Router (TWSR) and we call the network with such

nodes TWSRN. The circuits established between TWSRs are called time slot based lightpaths or ts-lightpaths (TSLP).

A TWSR node structure is similar to that shown in Figure 2.3(a) with the intermediate wavelength plane optical switch being an Optical Time Division Space Switch (OTDSS). The architecture of OTDSS is shown in Figure 2.12(b). The N inputs of the OTDSS are connected to N N x 1 wavelength multiplexers, respectively. All incoming packets in different ports are aligned with the local time slot of the switch using optical input synchronizers (OIS). The optical selective splitter (OSS) selects the data from either the local access unit or the input port to enter the reconfigurable optical space switching network (ROSSN). The data destined for the TWSR enters the LDBU and is later distributed to the local stations. In ROSSN, connectivity between input-output ports is configured on a per time slot basis. The optical output synchronizer (OOS) eliminates the jitter caused by ROSSN. The Optical selective coupler (OSC) is used to add packets which are stored in the LDBU.

The network design problem in such a network is called the routing, wavelength and time slot assignment (RWTA) problem. Given a network topology and traffic matrix, the objective is to identify the routing, wavelength assignment and the time slot assignment such that throughput is maximized. The RWTA problem with two simple heuristics - first fit and least loaded heuristics were proposed in [123]. The authors of [106] addressed the offline multi rate connection scheduling problem in ring topologies. The work in [110] provides an analytical model to characterize blocking performance of dual rate connections. The blocking characteristics of a TDM/WDM network with limited node reconfigurability is studied in [102]. The authors conclude that with only a limited reconfigurability at each node, performance similar to full reconfigurability can be achieved.

2.1.4   Optical Burst Switching

In Optical Burst Switching (OBS), information is transported in data units called bursts [88, 114, 28, 139]. An OBS network consists of user clients connecting to an optical cloud of interconnected OBS nodes. The client is typically an IP router with an OBS interface. The

38

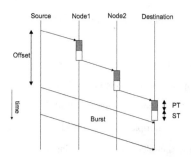

Figure 2.13   The burst is transmitted into the optical network at some time
offset after the control packet is sent to make reservations. PT
= processing time at a node, ST = switching time of an OXC.

core OBS nodes consists of an optical switching matrix, a switch control unit and routing and
signalling processors. For a good introduction on OBS, readers are referred to [13], a summary
of which is presented in this section.

The principal feature of OBS is the decoupling of data and control. At the ingress OBS
node, packets are aggregated into data bursts (DBs), routed through the network and disas-
sembled back into original packets at the egress OBS node. Each ingress/egress node maintains
a per-destination buffer called Burst Assembler (BA) which aggregates flows headed for the
same destination. For every DB, a Burst Header Packet (BHP) is created for control purposes.
Each DB is associated with three parameters - maximum burst size, minimum burst size and
maximum duration of a burst assembly process. The timer and size restrictions are imposed
since long bursts may lead to inordinate delays and higher burst losses while short bursts may
lead to more generation of control packets and wasted bandwidth.

Prior to the transmission of DB, a BHP is transmitted through the network which serves
as a channel set up request message. The BHP is transmitted on a channel different from DB.
A signaling protocol is then used in the network to communicate the bandwidth requirements
to carry the burst. The signaling protocol could either utilize a distributed one way signaling
procedure or a centralized end to end reservation procedure.

2.1.4.1 Signaling

The centralized method is termed as the wavelength routed OBS (WR-OBS) approach. In this approach, there exists a scheduler which has a global knowledge of the state of all the network switches and wavelength availability along all links. The job of this scheduler includes processing of incoming setup requests (BHPs), determination of routes, assignment of wavelengths, and constant monitoring of network usage. When the required resources are allocated, a positive ACK is sent to the source upon receipt of which the DB is transmitted.

The more common approach is the distributed one-way signaling mechanism. In stead of waiting for an acknowledgement for the BHP, DB follows BHP after an offset time and cuts through the nodes along the path. The offset time allows the control packet to reserve the required resources along the transmission path before the burst arrives as shown in Figure 2.13. The offset time is crucial since if it is incorrectly estimated, crossconnects in the path may not be configured and hence may lead to dropping of bursts. Several variants of OBS have been proposed in the literature which mainly differ in the way the offset time is calculated.

2.1.4.2 Offsets

Fixed Offsets:

The offset time is fixed and is equal to the sum of the total processing time at all the intermediate OBS hops and the switch reconfiguration time at the egress OBS node. Two different protocols have been proposed based on when the bandwidth reservation is actually done.

Tell and Go: Let t_o be the time, when the control packet, after being processed, tries to reserve the bandwidth at a node. The required bandwidth is reserved on the node starting at time t_o and is held until after the burst has been transmitted.

Just Enough Time: The required bandwidth is reserved on the node starting at time $t_o + T$ where T is the offset time. It also makes provision for possible buffering of DB at a local node so as to lower burst dropping probability.

Statistical Offsets:

The work in [118] proposes a scheme where an ingress node generates transmission tokens, based on a Poisson process with a predetermined rate of arrival. After PBH has been transmitted, DB waits until a transmission token has been obtained.

WR-OBS offsets:

The offset in this case is calculated to be the sum of three different components : (a) the time it takes a client to request resources from the centralized scheduler (b) the computation time of the resource allocation algorithm and (c) path signaling time.

2.1.4.3 Burst contention

Since DB does not wait for an ACK, it is not aware if the resources are really available in the network. A DB may arrive at a node and find that all the resources are busy. This may lead to a burst being dropped. In a variant approach, two ideas are specified : segment-first or deflect-first. The burst is divided into multiple segments. In the former, the lengths of the currently scheduled burst and the new contending burst are compared. The shorter one is segmented and its tail is deflected. In the latter approach, the contending burst is deflected if an alternate port is free. However, if it is not free, segment-first policy is applied on this port.

The above mentioned OBS techniques assume that the burst has been completely assembled before PBH is sent. It is also possible to send the control packet prior to assembling the entire burst. While this method reduces latency, it may lead to poor resource utilization since the control packet may not contain information regarding the length of the burst.

2.1.4.4 Routing

The routing of the burst through an OBS network can be done in one of the following ways (a) hop by hop basis using a table lookup algorithm to identify the next hop (b) map control packets to FECs and use MPLS for label swapping (c) Explicitly precalculated set up connections established through RSVP-TE or CR-LDP.

Though OBS has been researched well as a candidate for next generation optical network, there exists some concerns related to its feasibility for commercial deployment. The author

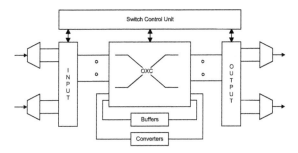

Figure 2.14   The functional blocks of an optical packet switch

of [90] observes that it may be easier to characterize the burstiness of the traffic and over-provision bandwidth rather than add an additional complex statistical multiplexing layer. The basic problem according to [90] is that at speeds over and above 40 Gbps, the physical impairments are difficult to handle even for a static optical network and hence it is not clear how dynamic networks like OBS can scale up to higher speeds.

2.1.5   Optical Flow Switching

In Optical Flow Switching (OFS), users request service for a duration of 100 msec or more [26]. The idea of flow switching was first introduced in [43]. To minimize network management and switch complexity in the network, flows are serviced as indivisible entities which means that data cells of the same flow follow the same network path and same wavelength. This is in contrast to OPS which breaks up a transaction into multiple cells and has the cells switched independently. OFS differs from OBS in that OFS is a scheduled flow-based transport architecture, thereby not requiring TCP for congestion and flow control [26]. A lightweight transport layer could be used instead to reduce cost. OBS, on the other hand, employs random access strategies, and is characterized by high blocking probabilities. This also implies that resources that are used to carry burst that are eventually discarded are wasted in OBS. The authors of [26] compare OFS and OBS and show that OFS can achieve better utilization and blocking performance at the expense of delay while for OBS, delay performance is poor owing to its collision and re-transmission strategy.

Another comparative study of OFS and OBS is reported in [127]. In this work, when a packet arrives at an ingress router, it checks if there is an existing flow to which it belongs. If there exists such a flow, the incoming packet is either buffered along with the other packets of the flow waiting in the queue or transmitted over the established circuit. Otherwise, a set up request is sent to establish a new circuit. If the circuit is not established, the flow is dropped. The ingress router considers that a flow has ended if there is no packet destined for a particular egress router for a pre-specified interval since the last packet of this flow arrived. The study concludes that OBS and OFS compete closely in terms of performance metrics like end to end delay and are sensitive to the access protocol parameters. A prototype OFS architecture is demonstrated in [44].

2.1.6 Optical Packet Switching

In a network that employs Optical Packet Switching (OPS), the smallest unit of transported information is a packet. Each packet consists of a header and a payload. There are two principal approaches to packet switching: (a) slotted networks (b) unslotted networks. A slotted network uses fixed sized packets and requires fragmentation and reassembly to support IP datagrams. It requires an input interface that synchronizes and aligns the packets against a local clock. An unslotted OPS network uses variable length optical packets. Such asynchronous networks are more flexible and robust than slotted synchronous networks. Most research work in the literature assume fixed size packets though a wide variety of switch architectures have been proposed for both fixed size and variable size packets.

Figure 2.14 shows the block diagram of a typical OPS node architecture. The architecture consists of three main sub blocks: an input interface, a switching core and an output interface. The input interface extracts the header and forwards it to the switch control unit for processing. The header may be processed either in the optical domain or in the electronic domain. Due to the lack of fast optical bit-level processing technology, the electronic approach is more practical at the moment. Based on the header information, the switch control unit determines

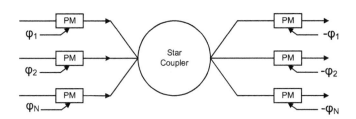

Figure 2.15   An implementation of an optical cdma network

an appropriate output port and a wavelength and configures the switch fabric to route the packet accordingly. It also generates a new header for the packet and forwards it to the output interface. When the packet is routed by the switching fabric, it may face contention. Contention is resolved in three dimensions: space (using deflection routing), time (using delay lines), wavelength (using converters). When contention is resolved, the packet is switched out to the right port where it is combined with the new header on the outgoing fiber link to the next node in its path.

OPS [134, 130] is still in its infancy and requires significant advancements in five different technologies for its practical realization. This includes optical switching, optical buffering, all optical wavelength conversion, packet synchronization and optical header processing. In optical switching, promising technologies include SOA based and Lithium Niobate based switches. In wavelength conversion, approaches based on cross gain modulation or cross phase modulation and wave mixing techniques are gaining importance. The slot level synchronization is researched using a cascade of delay lines and optical switches or through taking advantage of wavelength selective dispersive effects of fibers. There are two main approaches to formatting packet header. In the bit serial method, the header is transmitted on the same wavelength while in the subcarrier multiplexing method, the header is transmitted at a slightly higher frequency than the payload. An area of significant progress is in all optical label swapping, which refers to the techniques used to extract and replace the header without going through OEO conversion.

2.1.7 Optical Code Division Multiplexing

Optical Code Division Multiplexing (OCDM) differs from the other optical technologies in that it does not involve a switching fabric that needs to be reconfigured for grooming sub-wavelength level connections. It uses the principle of orthogonality of codes to allow multiple sub-wavelength connections to share a wavelength channel. However, code division multiplexing in the context of optical networks is different from that of wireless networks in that optical fiber is an intensity medium and binary signals are transmitted as pulses of light. An orthogonal optical code set is a set of code words, each of length N, that satisfies certain autocorrelation and crosscorrelation constraints mentioned in [96]. The individual characters of the codeword is called the chip and each codeword represents a bit. The number of 1s in the codeword of a codeset is called the Hamming weight. In ON-OFF keyed optical CDMA networks, the presence of a codeword signifies a 1 bit and absence of it signifies a 0 bit.

Codes are designed to be pseudo-orthogonal [96] to minimize interference. This allows multiple users to transmit simultaneously. When a receiver is tuned to a particular code word, it looks for power levels in certain chip positions. If the number of 1 chips received is above a certain threshold, it assumes that 1 bit has been sent. Else, it assumes that 0 bit has been sent. The threshold is typically set to the weight of the code. If there are enough codewords on the line such that there are at least threshold number of 1s in the appropriate 1 chip locations of the codeword under consideration, an error may occur. In this case, a 0 data bit may be misconstrued to be a 1 data bit. This type of error called a false positive and is the main cause of throughput degradation in CDMA networks.

The spectral efficiency of a CDMA network is given by A/N, where A is the number of active users. This is a measure of how the specified coding scheme compares with a pure TDM scheme. Consider a CDMA network with three users each operating at 40 Mbps. For zero false positive errors [67], three codewords, each of length 25, can be used. Since each bit is represented by 25 chips, each user effectively needs a chipping rate of 1 Gbps. The spectral efficiency of such a network is 3/25 = 12 %. Despite the low spectral efficiency, OCDMA may be a preferred technology because it allows asynchronous transmission without any media access delay.

An optical CDMA system can be implemented as shown in Figure 2.15. The system consists of the following [108]:

(a) optical phase modulators for spreading the spectra of incoming signals (b) couplers for combining and distributing input signals and (c) optical phase modulators for despreading the distributed optical signals.

To spread and despread the signals, a set of random functions that have no mutual correlation are prepared. The modulator on the input ports, modulates the input signal using the random function and is later demodulated at the output using the negation of the corresponding random function.

The biggest problem with optical CDMA is that existing technologies are unable to provide a large enough code space to accommodate increasing user base [90]. When a signal at 10 Gbps is encoded at 100 GHz, transmission impairments at the encoded rate are more than at the native rate.

2.2   Electronic Grooming

Electronic grooming techniques refer to the techniques built into the client layer that are overlaid on top of a grooming optical layer to carry traffic efficiently. A network that employs electronic grooming capabilities has at least one client node that processes traffic that is neither sourced or sunk by it and is called an electronic grooming network. If there exists no client node that processes transit traffic, then the network is called a pure optical grooming network.

The electronic grooming problem is a special instance of the virtual topology design problem with the assumption that the incoming traffic requests are subwavelength in nature. To understand the electronic grooming problem, we first introduce the routing and wavelength assignment subproblem.

2.2.1 The Routing and Wavelength Assignment Problem

Given a set of circuits, the problem of setting up the circuits by routing and assigning wavelength to each circuit is called the Routing and Wavelength Assignment (RWA) problem. From graph theory perspective, an optical network is viewed as a graph G(V,E) where V, the vertex set, refers to the nodes in the network and E, the edge set, refers to the links in the network. For each circuit (i,j), a path from i to j is assigned and a wavelength is allocated subject to the constraint that if the path traversed by the two connections share the same link, they are reserved different wavelengths. The RWA problem, in its pure form, assumes that wavelength conversion is absent in the network and is typically solved with an objective to minimize the number of wavelengths used.

If G is a path, star, spider or a tree network, the routing problem is solved since end nodes are connected by unique paths. For a path network, the wavelength assignment is equivalent to an interval graph coloring problem and can be solved tractably by a greedy algorithm [56]. If G is a star, wavelength assignment is equivalent to finding a minimum edge coloring in a bipartite graph, which is solvable efficiently combining Hall and Konig's theorems. For a spider network, [125] describes a method how wavelength assignment can be done in polynomial time. If G is a tree however, even in special cases such as binary trees, wavelength assignment is known to be NP-Hard.

If G is a ring, a connection can be routed in two different ways, and so both routing and wavelength assignment problems need to be addressed. For general topologies, this problem is proved to be NP-Hard in [125] by reducing it to an instance of a coloring problem. However, for tree networks, there is a polynomial time algorithm that yields a wavelength assignment solution that is no more than 5/2 optimal. For a ring network, it is possible to find a wavelength assignment solution that is no more than twice optimal. The RWA problem for any network becomes easier when wavelength conversion is allowed. The results in [91] shows that significant improvements can be achieved by using limited conversion. Also, multifiber networks have capabilities similar to networks with limited conversion capabilities and hence can lead to performance improvements.

2.2.2 The Electronic Traffic Grooming problem

For effective bandwidth utilization, optical networks allow several independent traffic streams to share the bandwidth of a circuit through a technique called electronic grooming or e-grooming. This is made possible by introducing some amount of electronic switching in the network. While this leads to transceiver and wavelength savings, system costs increase due to higher layer grooming electronic switches. Due to e-grooming, a connection may have to traverse multiple lightwave circuits before reaching the final destination. Such networks are called multi-hop networks.

E-grooming is done typically either through SONET or IP. SONET has stringent timing restrictions and can act as a client layer only for SHLP and OTDM based optical layer technologies. IP, on the other hand, is packet/label switched network and hence can act as the client for any of the WDM technologies mentioned in this chapter.

E-grooming in WDM has been researched extensively along several dimensions. Studied topologies, assumed traffic models, and analyzed cost functions have been the differentiating criteria amongst the various approaches. We discuss each of these dimensions in turn. For surveys on e-grooming in lightpath based networks, readers are referred to [11, 38, 59, 79]. The survey in [38] focuses on ring networks, while the survey in [11, 59, 79] discusses both ring and mesh networks.

The prime objective of e-grooming has been to increase resource utilization and decrease network costs. In this section, we study e-grooming in lightpath based networks.

2.2.2.1 Topology

E-grooming has been studied for various physical topologies like path, star, tree, ring and mesh networks [38]. Initial work was focused on ring based SONET networks and later diversified into mesh based IP networks. While a variety of architectures have been proposed for WDM rings, they primarily differ in their ability to exchange traffic across wavelengths. Some designs allow use of cross-connect at one or more nodes in the ring to allow traffic dynamics and for ADM savings while some others do not use cross-connects due to their excessive costs.

The mesh networks, on the other hand, differ in the extent of grooming capabilities available at the nodes. Nodes are equipped with source node grooming, partial grooming or full grooming (as described later) capabilities. In our study here, we classify the grooming problem into two - ring-based and mesh-based and review some important results in the next section. We describe some of the research issues that have been studied for peer, overlay and augmented models in the context of mesh networks. Our intent is not to be comprehensive but rather show some general trends in the approaches.

2.2.2.2 Traffic Model

Research work in the past has considered multiple types of traffic models: (a) static model, (b) dynamic model, (c) incremental model, (d) matrix set model (e) scheduled model (f) realistic model.

In the static traffic model, all the requests are known in advance and do not change. The general objective in such studies is to minimize cost while accepting all requests or maximizing throughput with a given set of resources. Since, response time is not a constraint in static provisioning, time consuming ILP formulations, or meta heuristics like simulated annealing, genetic algorithms, and tabu heuristics have been proposed.

In the dynamic model, calls arrive at a certain rate and depart after holding for a time interval with known distributions. Statistical models have been applied to analytically evaluate network performance in the presence of such traffic.

In the incremental traffic model, a call arrives dynamically but does not leave the system. The work in [1] conducted network planning across several years to produce a network that can carry all the traffic at the end of the planning horizon.

The authors of [109] use as input a set of different traffic matrices which is supposedly representative of time varying traffic as a new traffic model and provides a configuration solution with an objective to minimize resource consumption. The work in [103] also uses a set of traffic matrices to groom dynamic traffic in a WDM network.

In the sliding scheduled traffic model introduced in [119], the holding time of a connection is known in advance but the set up time is assumed to occur at any time in a pre-specified time window.

A poisson traffic model does not take into account the IP traffic elasticity and the interaction between the IP and optical layer and so a more realistic traffic model is proposed in [97]. The authors propose two realistic traffic models, and based on their simulations conclude that approximating IP traffic to be a CBR like traffic can lead to wrong conclusions when routing and grooming are considered.

2.2.2.3 Cost Function

The objective of the e-grooming problem is to optimize a cost function that is typically one of the following:

- Minimize equipment: add/drops, OXCs, fibers, transponders, wavelengths, time slots etc.

- Minimize the total amount of electronic conversion and total amount of traffic routed

- Minimize changes to existing topologies

- Minimize blocking probability

- Minimize number of physical hops or logical hops

- Maximize network utilization

- Minimize maximum number of circuit terminations

2.2.3 Electronic grooming in lightpath based architectures

2.2.3.1 Electronic grooming in Rings

Most of the initial e-grooming work was on rings since SONET over WDM was the predominant technology and SONET is configured in the form of rings. Each wavelength in SONET is operated at certain rates specified by $OC - M(M = 1, 3, 12, 48, 192, 788)$. The hierarchical multiplexing scheme in SONET allows an OC-M circuit to carry one or multiple $OC - N$ connections ($N <= M$). The ratio of M to the smallest N carried in the network is called the grooming ratio. SONET Add/Drop Multiplexers (ADMs) were used by nodes on the ring to add or drop channels to the wavelength.

In traditional SONET, one ADM per node per wavelength was required. With the introduction of OADMs, ADMs were required at a node for a wavelength only if the incoming wavelength does not optically bypass the node.

Most of the initial work in rings attempts to minimize the number of wavelengths required or the amount of blocking for the given set of lightpaths. The primary objective was to minimize the number of concentric SONET rings required to carry a given amount of traffic. However, the authors of [45] claim that this is not always the desirable goal due to the following reasons. In the near future, many WDM systems will be under-deployed since the bandwidth requirements may not scale up to the existing bandwidth per fiber. A more important goal for service providers would be to minimize the entire cost of the system. The dominant cost of the entire system is the electronics part (ADMs) and not the optics itself. With the above premises, it was suggested that minimizing the number of ADMs may prove to be a more useful objective. The static e-grooming problem is defined as follows. Given a traffic demand of low-rate circuits between pairs of nodes, assign traffic to wavelengths so as to minimize the number of ADMs used in the network.

Consider an example ring network 1-2-3-4-1 discussed in [79]. Suppose, each node pair has two OC-3 requests between them, and the line speed is OC-12. The request assignment to each ring is as follows:

$Ring1 : R_{1,2}, R_{3,4}; Ring2 : R_{1,3}, R_{2,4}; Ring3 : R_{1,4}, R_{2,3}$

It is seen that the specified assignment needs 12 ADMs. However, if assignment were done differently as follows:

$Ring1 : R_{1,2}, R_{1,3}; Ring2 : R_{2,3}, R_{2,4}; Ring3 : R_{1,4}, R_{3,4}$

The number of ADMs required would only be 9 which is a considerable improvement over the previous solution and results in a cost effective network.

The work in [29] proves that the general e-grooming problem is NP-Hard by considering an egress traffic model in a ring network and constructing a reduction from the bin packing problem. However, for special cases like uniform traffic in an egress model and in the hub with cross-connect model, optimal algorithms are described. For the more general all-to-all uniform traffic and distance dependent traffic, heuristics with near optimal solutions are discussed.

The work in [101] quantifies the maximum terminal equipment savings attainable using OADMs for rings carrying uniform all-to-all traffic and distance-dependent traffic. The objective is expressed as a "through-to-total" ratio of connections at each node in the network. The larger the ratio, the more the optical bypass and better savings are attained. It establishes that, over regions of interest, ADM savings increases with network size and internode demand.

The authors of [141] propose a cycle construction algorithm that can be used for both uniform and non-uniform traffic in unidirectional and bi-directional rings with an arbitrary grooming factor. In the first part, the designed heuristic packs the demands into circles, where each circle has capacity equal to the tributary rate and contains non-overlapping demands. The second part groups circles into wavelengths. The number of ADMs needed for a particular wavelength equals the number of end nodes involved. An end node is a node that terminates a connection in the circle. To minimize the number of ADMs, the heuristic attempts to match as many end nodes as possible when grouping the circles.

The traffic grooming problem has been studied in [37] under general traffic patterns with an objective to minimize switching costs. In the suggested approach, a ring is decomposed into a set of independent path segments and an ILP is solved exactly for each segment after which they are combined appropriately to yield the upper and lower bounds for the ring. The ILP for the path network do not have the wavelength assignment constraints and hence can be solved much faster than that for the ring. As the length of the path segment increases, the quality of the bound improves, and hence provably near-optimal solutions to large ring networks are obtained.

An integer linear program has been proposed for single-hop and multi-hop connections in [121] for arbitrary traffic. In the single-hop case, there is no traffic switching across wavelengths. In certain networks, a hub-node may be equipped with a Digital crossconnect (DXC), that terminates signals in the electronic form and perform time and wavelength switching. This method of grooming is called multi-hop grooming. A simulated annealing based on the cycle construction approach in [120] is proposed for the single-hop case and a greedy approach is adopted for the hub-node case for non-uniform traffic in unidirectional and bidirectional rings. The authors conclude that a multi-hop approach can achieve better equipment savings when the grooming ratio is moderate, but at the cost of more bandwidth. If, however, the grooming ratio is small, the single-hop approach tends to use fewer ADMs.

All the work discussed earlier focused on static traffic. In [15], a new class of traffic called t-allowable is defined which allows each node to source upto t circuits. These t circuits can be destined to any of the nodes in the network and the destination can be changed dynamically. The problem is formulated as a bipartite graph matching problem, and algorithms are developed to minimize the number of ADMs at each node.

Another approach to accommodate dynamic traffic is to use cross-connects. Cross-connects allow for traffic dynamics and reduces number of ADMs required. In [29], it was shown that any grooming that does not use a cross-connect can be transformed into one that uses a cross-connect without any additional ADMs. In [46], it was claimed that the cost savings can be as much as 37.5 %. The work in [47] introduces six node architectures with varying cross-connect capabilities on each node and deals with the topology design and connection routing sub problems. It identifies various scenarios under which one architecture would outperform the others in terms of cost metrics like average number of wavelengths, transceivers and hops traversed. The conclusion of the paper is that use of cross-connects adds flexibility to the network and hence provisioning can be done without knowing exact traffic requirements in advance.

While most of the work above dealt with grooming as a whole, the work in [45] deals with the wavelength assignment subproblem alone. The authors explain that wavelength conversion does not help achieve better savings and describe how lightpath splitting, rerouting and careful wavelength assignment may help improve ADM savings.

2.2.3.2 Electronic grooming in Mesh - Peer Model

With IP layer gaining in popularity as the client layer to the optical network, there is an increasing trend in migrating from ring topologies towards mesh topologies. The work in [146]

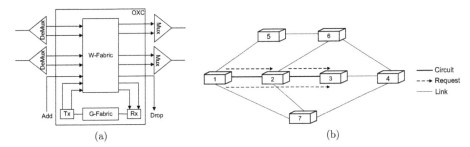

Figure 2.16   (a) The WGXC architecture that switches at wavelength level
through the W-Fabric and at a finer level through the G-Fabric
(b) An example e-grooming network

describes a mesh crossconnect architecture with e-grooming capability. The multihop partial
wavelength grooming OXC called WGXC shown in Figure 2.16(a) consists of two switching
fabrics - the wavelength fabric (W-Fabric) and the grooming fabric (G-Fabric). The W-Fabric
is capable of switching at the wavelength level granularity. However, it can also switch some
of the wavelengths to the G-fabric ports. Here, the wavelengths are regenerated, switched at
a fine granularity and combined along with the local add/drop signals. The resultant signals
are remodulated and routed into the W-Fabric to be switched out to one of the outgoing links.
The switch is partial grooming because only a few (but any) of the input signals can be routed
down to the G-Fabric which is determined by the number of ports supported on the G-Fabric.
If the number of G-fabric ports is equal to the total number of wavelengths on all the incoming
fibers, the WGXC is called a full grooming OXC. Full grooming OXCs may be implemented
opaquely and if the wavelengths are time slotted, the fabric switching speed is required to be
twice the line speed.

The basic technique behind e-grooming in mesh networks is as follows. Consider the network shown in figure 2.16(b). Three requests $R_{1,2}$, $R_{1,3}$, and $R_{2,3}$ arrive at the network. In
LP, three circuits $(1,2)$, $(2,3)$ and $(1,2,3)$ and two wavelengths need to be set up to carry
this traffic. However, in a network that supports e-grooming, only two circuits $(1,2)$ and $(2,3)$

are set up. Requests $R_{1,2}$ and $R_{1,3}$ are multiplexed onto one wavelength and transmitted on circuit $(1,2)$. Node 2 sinks request $R_{1,2}$, multiplexes $R_{2,3}$ and $R_{1,3}$ onto one wavelength and transmits it on circuit $(2,3)$. Thus, only one wavelength is required on each link. The majority of the research work is to identify the nodes where each request should be terminated before it reaches the destination, how each lightpath is routed and what wavelengths are assigned to each of the lightpaths.

Research on e-grooming mechanisms in IP over WDM networks has focused primarily on the peer model. Broadly, the work can be classified into two: analytical modeling and network design. We present some of the techniques that have been used in the literature for network design. The approaches can be classified as ILP-based for static traffic, and auxiliary graph-, network flows-, matrix-, and clustering-based techniques for dynamic traffic. Some of the studies mentioned below are independent of the client layer and hence is not constrained to IP being the higher layer.

Analytical approach

The work in [111] considers network nodes of two types: Wavelength Selective Crossconnect (WSXC) and Wavelength Grooming Crossconnect (WGXC) nodes. WSXC can switch traffic from one port to another but cannot switch streams between wavelengths. WGXC has the capability to switch signals in across fibers, time slots and across wavelengths. A network with only WSXC nodes are called constrained grooming networks while a network with some WGXC nodes are called sparse grooming networks. Analytical models with link independence assumption and capacity correlation assumption were designed to study the constrained grooming networks. The performance of networks with limited number of WGXC nodes were also modeled. The paper concludes that sparse grooming offers an order of magnitude decrease in blocking probability for high line-speed connections and multiple orders of magnitude decrease in blocking for low line-speed connections.

An analytical model to evaluate the blocking performance of grooming networks with heterogeneous grooming capabilities is discussed in [105]. The grooming network is modeled as a Trunk Switched Network which is a two-level network model in which every link in the network is viewed as multiple channels. Models with and without precise knowledge of the grooming architectures are considered and are observed to have similar performance.

Performance analysis of traffic grooming in mesh networks has been studied [128] in the presence of multi-granularity, multi-class, and multi-hop traffic. The single wavelength link is modeled using a queueing system based on continuous time Markov chains. Precise representation of the state will include the specification of traffic type in service, the number of client calls and the individual call bandwidths which may lead to state space explosion. Using the bulk arrival concept, the arrival of one client call is converted into the arrival of multiple fictitious micro calls, each having unit bandwidth and same service time. This reduces the state space to specify only the type of call and the number of micro-calls in service leading to a more tractable but approximate model. The link independence model is assumed and the blocking performance of both multi-hop and single-hop grooming were studied. The simulation results were found to be a good approximation to observations based on simulations. A similar model was studied in [133] to model sparse grooming networks.

It is observed in [112] that call requests that ask for capacity nearer to the full wavelength capacity are bound to experience higher blocking than those for a smaller fraction. This difference in loss performance is more pronounced as the traffic switching capability of the network is increased. The paper suggests that connection admission control mechanisms can be used along with wavelength assignment schemes to improve the fairness among connections with different capacities at the cost of increase in the overall blocking performance.

ILP based techniques

The work in [149] studies the static grooming problem with an objective to maximize network throughput. An ILP based mathematical formulation is presented for single-hop and multi-hop grooming for multigranularity connection with non-bifurcation constraints. Two heuristics with one that maximizes single-hop traffic (MST) and the other that maximizes resource utilization (MRU) are presented. Simulations were performed to observe the throughput with limited number of transceivers and wavelengths and were compared with the optimal solution. The paper concludes that MRU performs better if tunable transceivers are used and

MST performs better if fixed transceivers are used.

A generic graph model for grooming static traffic in a heterogenous grooming network environment is presented in [146]. The algorithm takes into account the heterogeneities in the network in terms of wavelengths, transceivers, conversion and grooming capabilities. Besides, it solves the grooming problem in a combined way in stead of splitting it into multiple sub problems and solving them independently. Three different policies were introduced, edge weight assignment principles were discussed and three traffic selection schemes were analyzed.

Auxiliary graph based techniques

Online approaches for provisioning connections of different bandwidth granularities were dealt with in [148]. For a connection to be established between two nodes, an attempt is first made to use an existing lightpath and if that fails to use a series of lightpaths. If the connection has not been accommodated yet, a new direct lightpath is set up or a mix of old and new lightpaths are used.

The authors in [33] propose a simple model for routing in peer model by assigning different weights to already existing circuits and new wavelength links. The special emphasis in the paper is on the signaling and protocol implementation aspects of the grooming scheme.

A generic graph model for grooming traffic in a heterogeneous grooming network environment is presented in [145]. The algorithm takes into account the heterogeneities in the network in terms of wavelengths, transceivers, conversion and grooming capabilities. Besides, it solves the grooming problem in a combined way in stead of splitting it into multiple sub problems and solving them independently. Three different policies were introduced, edge weight assignment principles were discussed and three traffic selection schemes were analyzed. The basic model can be used for static traffic as well [144].

In [2], two route selection strategies are studied for the peer model. Both identify multiple alternate routes between a given source destination pair. One strategy always routes a new request along the most loaded route while the other strategy tries to balance the load along multiple routes. The work in [136] also proposes two strategies based on the layered graph approach - channel level balance (CLB) and link level balance (LLB) and show that

LLB is better than CLB in most cases. The authors investigate the effects of wavelength conversion by studying multi fiber networks and conclude that when connection granularity is much smaller than wavelength granularity, wavelength conversion may lead to deterioration in blocking performance unless a connection admission strategy is used.

Dynamic routing in the peer model with inaccurate link state information is studied in [76]. The consistency and completeness of routing information is vital to achieve improved network throughput. However, the wavelength and bandwidth information may not always be accurate. The objective of this paper is to minimize the set up failures and blocking probability that results due to partial information. A probabilistic mechanism is applied to model the uncertainty of link state parameters and a cost function that takes into account this uncertainty is used to compute the route. The authors conclude that with their algorithms the impact of inaccurate link state information is significantly reduced.

A protection scheme to dynamically allocate paths for the peer model is designed in [143]. While the physical link is assigned a unit cost, a logical link is given a cost that equals the number of physical links traversed by it. It introduces a control parameter that determines the relative preference of physical links and logical links during route selection. The analysis of the control parameter suggests that the treating a logical link as no different from a concatenation of physical links yields the best results.

Network flow based techniques

The authors of [73] study the problem of traffic grooming to reduce the number of transceivers in optical networks. This problem is shown to be equivalent to a certain traffic maximization problem. An ILP formulation is presented and a greedy heuristic that uses the min cost flow problem is described. Simulation and ILP results were compared for uniform and random traffic pattern for small networks.

An algorithm for integrated routing for the peer model was presented in [70]. It uses a graph based approach that contains both the virtual and physical links. The model identifies all the min cuts for every possible ingress-egress pair and considers a link to be critical for this pair, if this link appears in at least one of its cuts. Each link is assigned a cost based on the number of LSR pairs for which this link is considered critical. By discouraging a new flow from using these links, the amount of residual capacity in the network can be maximized at every iteration. However, the complexity of this heuristic is high since it has to compute max flow for all node pairs.

Matrix based techniques

The work in [104] models a WDM grooming network with nodes employing heterogenous switching architecture as a Trunk Switched Network (TSN). A TSN is a two level network model in which every link in the network is viewed as a set of channels. In this model, the control plane information like link properties are provided in the form of a matrix representation. By defining a generalized version of matrix multiplication based on some chosen operators, link information is combined to yield path information allowing path selection, sub trunk assignment and channel assignment for a traffic request.

Clustering based techniques

The study in [142] uses a clustering technique called Blocking Island paradigm (BI) to propose an integrated grooming algorithm in peer model. BI provides an efficient way of abstracting resource availability in a communication network. BI clusters nodes in the network according to bandwidth availability. One of the distinct features of this paradigm is the ability to identify the existence of a route of sufficient capacity without having to compute a route based on shortest path algorithms. Since only a small segment or island of the network is studied, it is fast and scalable and yields better solutions than provided in [70].

A framework for hierarchical traffic grooming based on a clustering approach is presented in [27]. The authors of this paper decompose a network into multiple clusters and select a hub node which will act as the grooming hub for the traffic originating and terminating at local nodes. At the second level of the hierarchy, the hub nodes form a virtual cluster for the purpose of grooming intra-cluster traffic. While the work presented in [27] assumed that the clusters were already given, the work in [36] suggests a mechanism to choose clusters based on the K-center problem.

2.2.3.3 Electronic grooming in Mesh - Overlay Model

The authors of [135] propose a grooming heuristic for the overlay model that puts a constraint on the maximum number of hops allowed in routing a connection. Through analysis and simulations results, they show that there exists an optimal hop constraint for each network configuration that leads to efficient capacity utilization.

The work in [137] focuses on designing a LSP-level shared and dedicated partial protection scheme for the overlay model. It allows for single hop primary paths but up to two hop shared backup path. It concludes that a combined IP and WDM layer protection scheme is more resource efficient than WDM level shared protection. It studies the blocking performance as a function of LSR speeds and concludes that high router capacities will be required to support systems with large number of wavelengths.

The authors of [74] provide a MILP formulation for protection only in the IP/MPLS layer for the static overlay model. By designing a simple heuristic, they conclude that a scheme that protects from both LSR failures and WDM single link failures costs only marginally more than a scheme that provides protection from LSR failures alone.

The problem of survivable network realization in the context of overlay model was reported in [94]. The authors of [94] introduce three new lightpath services (Disjoint-Pair, Disjoint-Triangle and Disjoint-Triple), which when combined with three other lightpath services (Protected, Unprotected, and Dual-Homed) defined in [98], can be used to design cost effective survivable networks. The reported results suggest that for a specific network scenario, the algorithm proposed in [94] achieves up to 18.8 % reduction in path costs and 24.6 % reduction in bandwidth costs over the results obtained in [98].

2.2.3.4 E-grooming in Mesh - Augmented Model

The most significant contribution of the work in [75] is to identify a specific type of control information that could be exchanged along the IP and optical networks for the augmented model. The paper suggests that the WDM layer pass L_{ij}, the number of lightpaths that can be established between LSRs i and j, to the IP/MPLS layer. L_{ij} could be the number of common free wavelengths available on every link of the path identified by the routing algorithm. It is approximated that the amount of capacity available between i and j is the sum of residual capacities on the existing logical topology and the amount that could be used in the future (L_{ij}). By assigning a cost to the link that is inversely proportional to the total residual capacity, the algorithm achieves an order of magnitude improvement in results than provided in [70].

2.2.4 Electronic grooming in other architectures

E-grooming has been studied primarily in lightpath based networks. The effect of e-grooming in OBS networks [100], LT networks [6, 138], LTR networks [60, 146] and DaC networks [40, 6] have also been investigated. The nomenclature that is followed here to refer to a network with electronic grooming capability is to suffix the architecture with the letter TG (Traffic Grooming). For instance, LP, SLT, DLT and LT networks when equipped with grooming capability in the client layer are called LP-TG, SLT-TG, DLT-TG and LT-TG networks.

Four kinds of networks based on the DaC switch architecture were studied in [40] - (1) networks with dropping and extension capabilites (2) networks with dropping only capabilities (3) networks with extension only capabilities (4) traditional lightpath networks. Electronic grooming algorithms were developed for these networks and were studied using a graph based approach. It concluded that DaC based networks performed better than lightpath based approaches when the transceivers at the nodes were limited.

The authors of [146] consider the dynamic multi-hop grooming problem using light trees and report that such architectures outperform LP networks significantly in terms of network throughput and resource efficiency. An auxiliary graph based algorithm to support multi-hop electronic grooming called dynamic tree grooming algorithm is proposed in [60]. When a request arrives at a network, it is accommodated in one of the three ways: (1) an intermediate non-leaf node on an existing light tree can be configured to receive the data (2) a new branch can be added to the tree (c) an existing branch can be extended linearly. In network scenarios with limited transceivers, light trees were shown to be superior to traditional lightpath based networks in [60].

The authors of [138, 140, 41] present a heuristic for dynamic provisioning in multi-hop light-trail networks. The work in [138] compares the multihop LT approach with multi-hop LP approach and show that for specific traffic scenarios e-groomed LT can outperform e-groomed LP networks.

2.3   The problem with electronic grooming

Electronic grooming results in improved wavelength utilization, but, it also results in the lack of bit rate, encoding, and protocol transparency [32]. For instance, there needs to be a unified upper layer that performs the grooming functionality. With transparency being lost, seamless and cost effective upgradability feature of optical networks is also lost. Even with e-grooming being implemented in the network, best utilization is possible only through peer model which appears to be practical only in the future.

Another big source of concern is the feasibility of high speed electronics that does the e-grooming functionality. This is because routers have to process the transit traffic apart from handling the traffic sourced/sunk by the local node. The cost per electrical port is more expensive than cost per optical port. While an LSR can be faster than a table look up based router, the incoming LSPs will still have to be stored, scheduled and label swapped before switching out to the output port. The time available for processing and scheduling an LSP is drastically low at high router speeds.

There is a growing gap between link and processor speeds [23]. With increasing line speeds, packets can arrive at a port at speeds exceeding the capability of a single processor. However, only packets of the same flow in the incoming stream have dependencies and hence the processing of independent flows can be distributed to several processors working in parallel.

With large router capacities, there is lots of output contention and hence fast arbitration is required. As mentioned in [23], for a 40 Gbps switch port with 40 byte packets, the arbitrator has only about 4 nanoseconds to resolve the contention. Buffer management and memory access speeds could prove to be a bottleneck in gigabit speed switches. A large router usually needs multiple racks with huge amounts of information being exchanged among the racks. The interconnect technology for the backplane is usually optical and is associated with high costs and excessive power consumption.

2.4   Summary

There is a wealth of literature in the field of grooming and this chapter gives a high level picture of the various approaches, tools and techniques that have been applied to study this problem. Grooming is required to cost-effectively multiplex sub-wavelength requests into high bandwidth wavelengths. The nature of grooming available in the optical layer typically depends on the granularity and timescales at which the switching fabric reconfiguration happens in the optical network. This is possible either through circuit switched mechanisms like waveband-, wavelength- and sub-wavelength level switching or through statistical mechanisms like burst-, flow- or packet switching. Sometimes, a grooming optical layer may be overlaid with electronic switching capabilities in the client layer which helps carry traffic more efficiently. Such a mechanism is called electronic grooming and is characterized by intermediate nodes in the network processing transit traffic. Though electronic grooming leads to efficient network utilization, it may lead to performance bottlenecks with the growing gap between link and processor speeds. There is an increasing trend towards achieving efficient network utilization purely through optical grooming for scalability purposes and resorting to electronic grooming only when it is absolutely required.

63

CHAPTER 3.   Network Design and Provisioning

There are two types of traffic that we consider in this chapter - static and dynamic. For static traffic, all the demands are known in advance. The path selection and resource assignment strategies could be optimized using ILPs or algorithms of high time complexity since computation time is not a serious constraint. Such strategies are studied under the topic of static network design. In the case of dynamic traffic, requests arrive in a particular sequence and has to be routed in the same sequence. In such a scenario, computation time available for resource allocation may be less and hence fast and efficient heuristics are required. The algorithms that are proposed for this are studied under the topic of dynamic connection provisioning [7, 8].

The main contribution of this chapter is in developing network design and connection provisioning algorithms for Path Level Aggregation of Traffic in metrO Optical Networks (PLATOONs). As explained in Chapter 2, there are four types of path level aggregation strategies - P2P, P2MP, MP2P and MP2MP. In this chapter, we consider only four architectures - LP, SLT, DLT and LT. We study and compare these different strategies by using an unified graph model.

This chapter is organized as follows. We propose the PLATOON architecture for metro networks and present a discussion on several possible crossconnect configurations for the PLATOON nodes. We study the static network design problem for single-hop PLATOONs and two-hop sparse e-grooming PLATOONs. Next, we design grooming heuristics for modeling multi-hop PLATOONs with dynamic traffic. Finally, we discuss network provisioning with constrained router speeds. The simulation results and performance evaluation studies based on these heuristics are presented in Chapter 4. Though our work described in this Chapter is

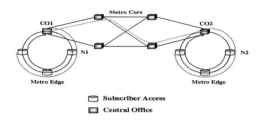

Figure 3.1   A metro network employing WDM solution. CWDM rings can
be deployed in the metro edge. DWDM rings or mesh are found
in the metro core.

generic, the discussion is presented in the context of LT networks (unless explicitly stated otherwise) and extensions to other types of networks like SLT, DLT and LP are straight forward.

3.1   PLATOON Network Architecture

We propose a two-tier PLATOON architecture for the metropolitan networks as shown in
Figure 3.1. The metro network consists of a set of rings with each ring having a hub node.
The hubs in turn are connected in the second tier to form a mesh network. The metro edge
traffic pattern is hubbed while in core networks, the traffic pattern is meshed. We propose
a CWDM light-trail bus/ring architecture for metro edge networks and a DWDM light-trail
ring/mesh architecture for metro core networks. The edge networks consist of reconfigurable
OADMs while the core networks consist of OXCs. The downstream trail is used for the hub
(central office) to transmit data to all the other nodes (access points) on the bus/ring and the
upstream trail is used for the access points to transmit data to the hub [17, 116].

For example, if node N1 in the figure has data to be sent to node N2, the data is first
sent on the upstream trail to the central office CO1 and then routed via the metro core which
then reaches node N2 via the downstream trail originated by CO2. Our research work focuses
on the network design issues in metro core mesh networks. Since the edge and core networks
employ different WDM technologies, the signals from the edge may have to be regenerated at
the hub node for adaptation purposes before being transported across the metro core. Our

current work is primarily focused on developing algorithms for design of the mesh based metro core networks.

3.2   Light-Trail Crossconnect Architecture

The work in [52] discusses only node architecture for single-fiber-in, single-fiber-out networks as shown in Figure 2.8(a). In a typical mesh setting, multiple fibers with each fiber having multiple wavelengths pass through a node and hence the architecture of the optical crossconnect becomes important. An architecture for light-trail networks is introduced in the context of multicasting in [51]. However, it again considers only fiber-level switching and not wavelength-level switching. In [5], we described the light-trial access unit (LAU) which comprises of the optical splitter, shutter and combiner through which a client can access the LT circuit with the help of a transmitter and a receiver.

In this section, we introduce different switch architectures and illustrate our motivation for different architectures using a simple example. Suppose that node A is active on trail t_1 and inactive on trail t_2. A design issue that needs to be addressed is whether t_2 needs to traverse the node A LAU or not. Having an optical switch that allows t_2 to bypass the node A LAU may seem to be the best thing to do to prevent t_2 from suffering unnecessary losses, but this leads to other transmission engineering issues. Namely, if t_2 bypasses node A LAU while t_1 traverses it, t_2 will have more signal strength than t_1 when both the signals exit node A. Signals of significantly different power levels may lead to amplification problems since one signal may saturate the EDFA because of its high power level while another may not get amplified much because of its low power level.

We could have two approaches to counter this problem. One solution is to let both t_1 and t_2 traverse the node A LAU so that they will have the same, but low power level. The second solution is to let t_2 bypass node A LAU through a switch, use a low gain amplifier like semiconductor optical amplifier (SOA) to compensate for the local losses t_1 suffers on node A LAU, so that when t_1 emerges out, it will have the same power level as t_2. In both approaches, finally when the span losses and DWDM component losses have accumulated, amplification by a high

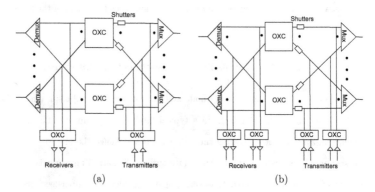

Figure 3.2   Configuration C1 (a) Tunable transponders (b) Fixed wavelength transponders

gain EDFA becomes possible since both the trails have the same power level. However, the first solution may require higher gain amplification or more amplifer/regeneration stages than the second. Based on the two approaches here, we propose different cross-connect configurations (C1, C2, and C3) and analyze their capabilities and hardware requirements.

3.2.1   Crossconnect Configurations

Let F be the number of fibers and W be the number of wavelengths per node.

Configuration C1: In the configuration shown in Figure 3.2(a), though the signals destined for the local node are received before they enter the $F \times F$ OXC, it is possible for the signals to be tapped at its exit ports as well. So, the above configuration is equivalent to having an LAU at each exit port of the $F \times F$ OXC. It requires F Mux, F Demux, W $F \times F$ OXCs, and WF LAUs are required. If tunable transponders are used as seen in Figure 3.2(a), K tunable transponders, a $WF \times K$ OXC, and a $K \times WF$ OXC are required. If fixed transponders are used as seen in Figure 3.2(b), K transponders per wavelength, W $F \times K$ OXC, and W $K \times F$ OXC as seen in Figure 3.2(b). We see that when $K = F$, the additional cross-connects are not required. Here, every signals entering a node has to go through an LAUs on this node. This

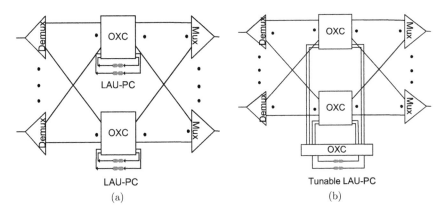

Figure 3.3   (a) Configuration 2 with fixed wavelength transponders (b) Configuration 3 with tunable transponders

ensures that every signal emerging out has the same, but low power level.

Configuration C2: The configuration shown in Figure 3.3 (a) allows K LAU-PCs (LAUs with power compensation) for every wavelength with K being a maximum of F. It requires F Mux, F Demux, W $(F+K) \times (F+K)$ OXCs, WK LAUs, WK SOAs, and K transponders per wavelength.

Configuration C3: The configuration shown in Figure 3.3 (b) consists of two levels of cross-connects, tunable lasers and broadband receivers. This allows support for K LAU-PCs for the entire node, with K being a maximum of F. This configuration requires F Mux, F Demux, W $(F+K) \times (F+K)$ OXCs, $(WF+K) \times (WF+K)$ OXC, and K LAU with tunable transponders, and K SOAs.

3.2.2   Architectural Tradeoffs

In C1, all signals go through LAUs, and hence suffer high losses, requiring high gain EDFAs. However, only few of the signals that are tapped at the LAUs are required by the higher layer. In C2 and C3, some signals can bypass the LAU-PC units by directly being switched from the demultiplexer port to the appropriate multiplexer port. The signals that need local processing are routed to one of the LAU-PC ports to be tapped. These packets get recirculated into the fabric to be switched back out on the right multiplexer port. Note that, the LAU-PC units can be used to process both light-trails and lightpaths. This allows lightpaths to interoperate with the light-trails. Recall that a lightpath is just a special case of a light-trail where in the end points alone have access to the channel. If the cross-connects are configured so that the wavelength bypasses all of the intermediate LAUs, the circuit that results would be a lightpath. This gives the network designer the opportunity to seamlessly integrate lightpaths and light-trails as per the network requirements.

The switch sizes required by C3 is significantly bigger than that of C2. Even C1 with tunable transponders requires switches of large dimensions. Large switches are harder to build since they need analog beam steering micromirrors whereas small switches can be realized using a variety of technologies [92]. In C2 and C3, SOAs compensate for the local insertion losses caused by the splitter, shutter and combiner. SOAs are noisy, expensive and prone to cross-talk. However, providing a gain just before the LAU can help the local receiver achieve a better dynamic range.

The number of LAUs required by C2 and C3 are much fewer than the requirement of C1. For instance, if W=64, F=4, and if the node is active only on 8 trails, C1 still has to provision 256 LAUs while just 8 per wavelength are required in C2 and 8 per node are required in C3.

In C2, though the OXCs are reconfigurable, but the transponders themselves are not. So, a decision needs to be made ahead of time regarding the number of transponders required for each wavelength. This may make the network planning constrained. It is expensive to have a transponder deployed and not used while the associated signal is being ignored by the higher layer on the node. These problems can be best avoided by making the transponders also reconfigurable as supported by C1 and C3. It should be noted that tunable transponders are more expensive than their non-tunable counterparts but the increase in premium is only modest [90]. The tunable components address two important problems. First, it may be difficult to forecast which wavelengths may have increased traffic since traffic may grow unpredictably. Secondly, the operational cost associated with stocking multiple part numbers to

address different wavelength requirements could be high. So, an analysis is required to justify
if the increased premium is worth the possible returns.

Configuration C1 has an advantage which the other configurations do not have. Consider
a linear network 1-2-3. When request $R_{1,3}$ arrives, a circuit {1,2,3} is established. The transmitter on node 1 and receiver on node 3 are active on this circuit. Now, suppose a request
$R_{2,3}$ arrives and there is sufficient residual bandwidth in the existing circuit to carry this call.
In configuration C1, the transmitter on C2 can simply tune into the circuit and this request
can be set up. The medium access protocol allows the two connections to share the circuit.
In configuration C2 and C3, this is possible if the circuit was initially routed through a LAU
on node 2. Otherwise, the node 2 crossconnect should be reconfigured to route it through one
of the LAUs and have a local laser tune into the circuit. Though, this kind of reconfiguration
may happen only when a new connection is multiplexed on an existing circuit, it temporarily
disrupts the circuit and hence may not be desirable. Despite this problem, this configuration
may be preferable since it allows interoperability between lightpaths and light-trails.

A detailed trade-off among these configurations will include issues related to capital expenses, operational expenses, expected traffic growth, ease of network deployment and management and is specific to a deployment scenario. For our current work, the configuration in
$C1$ with tunable transponders are used unless explicitly stated otherwise.

3.3   Single-Hop Static Network Design

In WDM networks, the client signals are likely to be of multiple granularities and have
sub-wavelength capacity requirements. In this section, we study the design of PLATOONs in
the presence of subwavelength traffic and limited network resources.

The work in [39] provides ILP formulation for the static LT network design problem but
does not take wavelength assignment into account and does not minimize the wavelength usage
or the communication equipment resources as we do here. The work in [18, 20] provide the ILP
formulation and heuristics for MP2P sharing along a path whereas LT corresponds to MP2MP
sharing strategy along a path. The heuristics presented in [50] are for ring networks and are

not applicable in the context of mesh networks that we discuss here.

In this section, we provide an ILP formulation for the static network design problem. We decompose the design problem into two subproblems - the trail routing subproblem and wavelength assignment subproblem. We define the trail routing problem, describe two variants of it and show that both variants are NP-Complete. We present three simple polynomial time heuristics to efficiently solve the single-hop light-trail design problem.

3.3.1   ILP Formulation

The light-trail routing and wavelength assignment problem can be defined as follows: Given a network graph G(V,E), a request matrix R(V×V), identify the trails required to carry all requests so as to minimize resource utilization. The objective is to minimize the optical layer and the electrical layer costs. The electrical layer costs include the aggregate number of transmitters and receivers required to carry the given traffic and the optical layer costs include the number of different wavelengths required in the network. By minimizing the number of communication equipment required to carry a given static traffic, more resources are left available in the network to handle the incremental and dynamic traffic efficiently. We introduce our notation below, provide the objective function and constraints and describe the formulation in this section.

N - number of nodes in the network (data)

C - capacity of a wavelength (data)

W - number of wavelengths on each link of capacity C (data)

S - maximum allowable trail size in the network (data)

LT - set of all possible light-trails in the network of size S or less (data)

LT_t - an instance of a light-trail $LT_t \in LT$ (data)

len_t - length of a light-trail LT_t (data)

LT_t^r - set of requests that can be supported by LT_t based only on the containment constraint (data)

$LT_t^{i,j}$ - assumes 1 if trail LT_t traverses link (i,j), 0 otherwise (data)

$s, d = 1..N$ number assigned to each node (index)

$R_{s,d}$ - traffic request between node s and node d, assumed to be sub-wavelength(data)

$t, t_1, t_2 = 1..\|LT\|$ - number assigned to each light-trail(index)

$w = 1..W$ - number assigned to each wavelength (index)

$X_t^{s,d}$ - assumes 1 if (s,d) is assigned to t, 0 otherwise (variable)

T_t - assumes 1 if trail t supports at least one node pair, 0 otherwise (variable)

L_t^w - assumes 1 if wavelength w is assigned to trail t, 0 otherwise (variable)

$W_t^{r,s}$ - assumes 1 if node s on trail t needs a receiver, 0 otherwise (variable)

$W_t^{x,s}$ - assumes 1 if node s on trail t needs a transmitter, 0 otherwise (variable)

U_t - number of transmitters and receivers required for the trail t (variable)

Objective :

$$Min \quad \sum_t U_t \tag{3.1}$$

Request Assignment Constraint :

$$\sum_t X_t^{s,d} - 1 = 0 \quad \forall s, d \in V \tag{3.2}$$

Capacity Constraint :

$$\sum_{(s,d) \in LT_t^r} R_{s,d} X_t^{s,d} - C \leq 0 \quad \forall t \tag{3.3}$$

Trail Assignment Constraint :

$$T_t - X_t^{s,d} \geq 0 \quad \forall (s,d) \in LT_t^r, \; \forall t \tag{3.4}$$

Wavelength Continuity Constraint :

$$\sum_w L_t^w - T_t = 0 \quad \forall t \tag{3.5}$$

Wavelength Assignment Constraint :

$$\sum_t L_t^w - 1 \leq 0 \quad \forall w, \{t : LT_t^{i,j} = 1\}, \forall (i,j) \in E \tag{3.6}$$

Receiver Usage Constraint :

$$W_t^{r,s} - X_t^{d,s} \geq 0 \quad \forall (d,s) \in LT_t^r, \ \forall t \tag{3.7}$$

Transmitter Usage Constraint :

$$W_t^{x,s} - X_t^{s,d} \geq 0 \quad \forall (s,d) \in LT_t^r, \ \forall t \tag{3.8}$$

Resources Usage Constraint :

$$U_t - \sum_{s \in LT_t} W_t^{r,s} - \sum_{s \in LT_t} W_t^{x,s} = 0 \quad \forall t \tag{3.9}$$

Variable Range Constraint :

$$X_t^{s,d}, T_t, L_t^w, W_t^{r,s}, W_t^{x,s} \in (0,1), U_t \in I \tag{3.10}$$

The electrical layer costs include the number of transmitters and receivers and the optical layer costs include the number of different wavelengths required in the network. By minimizing the number of communication equipment required to carry a given static traffic matrix, more resources are left available in the network to handle the incremental and dynamic traffic efficiently. For the Minimize Wavelength Links formulation, the objective function to be used is $\sum_t len_t T_t$. Alternatively, the objective may be to minimize both the optical layer and the electrical layer costs with more emphasis on the optical (electronic) costs. Towards this objective, we first fix the number of wavelengths (transceivers) and check if the ILP yields a solution. If a solution is not found, then larger number of wavelengths (transceivers) are attempted. Otherwise, fewer number of wavelengths (transceivers) are attempted. This procedure is iterated until the minimum number of wavelengths (transceivers) is found.

Wavelength continuity constraint ensures that every trail is given exactly one wavelength throughout its route and the wavelength assignment constraint prevents two trails traversing the same link from being assigned the same wavelength. Equation (3.7) accounts for the number of receivers required on a trail t and equation (3.8) accounts for the number of transmitters required on a trail t. Equation (3.9) keeps track of the total number of transmitters and receivers counted separately. Let T be the number of trails required to carry a traffic matrix R. Then, it can be seen that,

$$\lceil \frac{\sum_{s,d} R_{s,d}}{C} \rceil \leq T \leq \sum_{s,d} \lceil \frac{R_{s,d}}{C} \rceil$$

For $R_{s,d} << C$, the bounds are loose, but as $R_{s,d}$ approaches C, the bounds become tighter. The bounds become exact when $R_{s,d} = C \quad \forall s, d \in V$, in which case, the ILP reduces to the static lightpath RWA problem.

3.3.2 Trail routing problem description

The light-trail network design problem also called the Trail Routing And Wavelength assignment problem (TRAW) is a hard problem to solve. The solution to the problem includes identifying the set of trails, assigning connections to each trail and allocating a wavelength to each trail. In order to simplify the problem, it can be reduced to two subproblems - trail routing subproblem (TRP) and wavelength assignment (WA) subproblem. The wavelength assignment subproblem is proved to be NP-Complete in [30] through a reduction from a well known hard graph coloring problem. In this section, we define two variants of TRP and prove that both are NP-Complete.

The trail routing problem can be formally defined as follows: Given a network $G'(V, E)$ and a request matrix $R(V \times V)$:

Minimize Path Cost: if a cost function $f : f(e) \to c_e$ where $c_e \in \Re$ is defined for every edge $e \in E$, if the cost of a trail is the sum of the cost of its links and the cost of a solution is the sum of the cost of the trails, find the minimum cost solution required to carry R. If the cost function assigns a unit cost to all the links, the objective function minimizes the number of wavelengths links required to carry all the requests.

Minimize Transceiver Cost: if a cost function f is defined for every trail to be the cost of active equipment on the trail and the cost of a solution is the sum of the cost of the trails, find the minimum cost solution required to carry R. In our work, we assume the cost of transmitter and receiver to be equal and the cost of the trail to be the number of active transmitters and receivers on the trail.

3.3.2.1 Minimize Path Cost Formulation

Define the decision version of the optimization problem as, given, $TRP_{wl} = <G', R, f, k>$, identify if there is a trail assignment in G', to satisfy R, with a cost of at most k using cost function f. $TRP_{wl} \in NP$, since, for a certificate specifying the set of trails satisfying R, we can check in polynomial time, if the total cost of the trails is at most k and if the above three constraints are satisfied.

To prove that TRP_{wl} is NP-Hard, we introduce the problem called $HAM-PATH \in NPC$, and by using a reduction technique show that $HAM - PATH \leq_p TRP_{wl}$. $HAM - PATH$ is a decision problem that decides whether or not an input undirected graph has a hamiltonian path (a path between two vertices of a graph that visits each vertex exactly once). It can be proved that $HAM - PATH \in NPC$ based on results available in [34], however this proof is omitted here.

For a given undirected graph G, a polynomial time algorithm is run to check if the graph is connected. If the graph is not connected, it does not have a hamiltonian path. Otherwise, we construct a network G' as follows. For every vertex in G, introduce a node in G'. Let the number of nodes in G' be m. For every edge in G, introduce two directed links in G'. Assign a cost function f, which maps every edge in G' to a unit value. Define R such that diagonal elements of R are 0 and the rest are 1. This ensures that every node has a unit traffic to be sent to all other nodes in the network. Assign the wavelength to have capacity m^2. This allows a single trail to have a capacity greater than the sum of all request sizes. We now claim that the graph G has a hamiltonian path if and only if G' has a trail assignment of cost at most $k = 2m - 2$. To complete the proof, we show that this transformation is indeed a polynomial time reduction.

Suppose G has a hamiltonian path. G' also has the same hamiltonian path since the reduction preserves all the nodes and links in G. From Lemma 1, we show that a trail assignment

75

of cost $2m - 2$ can be found.

Suppose that G does not have a hamiltonian path. It can be proved easily that G' also does not have a hamiltonian path. We claim that a trail assignment of cost $2m - 2$ or less can never be found. For this, we establish a lower bound on the cost of any trail assignment using Lemma 3 (given below) to be $2m - 2$. In Lemma 4, we conclude that even if there is one trail in the solution set that is not hamiltonian, the cost of the solution is always greater than $2m - 2$. Since G' does not have a hamiltonian path, the cost of the solution is always greater than $2m - 2$. This completes the reduction.

Lemma 1: If G' has a hamiltonian path, there exists a solution of cost 2m-2

Suppose that G' has a hamiltonian path $T_1 = \{v_1, v_2, .., v_m\}$. Based on the reduction, $T_2 = \{v_m, v_{m-1}, .., v_1\}$ is also a valid hamiltonian path in G'. Since, the capacity of each trail is m^2, T_1 and T_2 can serve all the requests in their respective containment sets. This satisfies all the requests in R and thus, exactly two trails are required. Since the cost of each trail is $m - 1$, the total cost of the assignment is $2m - 2$.

Lemma 2: Each node has to appear in at least two different trails in any solution for the given problem

Consider a trail T and a node $i \in T$. Since T is a unidirectional simple path, there exists no single position in T, where a node can receive from and send to all the other nodes in the network. Since the traffic matrix R has a unit traffic to be sent from node i to every other node and a unit traffic to be received from every other node, i has to appear on at least two different trails to satisfy all it's requests.

Lemma 3: The minimum cost trail assignment for a network with m nodes is 2(#nodes−1).

This can be proved by induction on the number of nodes of the network. For a two node, two link network, this lemma is easy to see. Assume that the lemma is true for an arbitrary network G'_k with k nodes and the minimum cost assignment is $2k - 2$. Consider a network G'_{k+1} with $k + 1$ nodes and suppose an optimal trail assignment for G'_{k+1} costs less than $2k$. Also suppose that S is the solution set that consists of the list of trails that conforms to the three constraints and length of the set S is defined to be the sum of the number of links present in all the trails of S. We will see how this supposition can lead to a contradiction.

Transform G'_{k+1} into a new graph G^o_k such that for some arbitrary $i, s, t \in G'_{k+1}$, if $(s, i), (t, i) \in G'_{k+1}$, introduce $(s, t), (t, s) \in G^o_k$ and remove i from G^o_k. However, the cost function and traffic are preserved for G^o_k except for that of node i. For G^o_k, based on the transformation, if each occurrence of i is removed from S to form S', the resulting S' is a valid solution and can satisfy all the requests of the k nodes in G^o_k. Since, node i occurs in at least two trails of S (from Lemma 2), by removing i, the length of S' is at least two less than the length of S. Hence, the cost of assignment S' is at least two less than the cost of assignment S. Since S costs less than $2k$, S' should cost less than $2k - 2$ which is a contradiction to our assumption about the cost of k node networks. Hence, S costs at least $2k$. This completes the induction step.

Lemma 4: A solution that has at least one trail that is not hamiltonian strictly costs more than a solution that comprises of only trails that are hamiltonian.

Consider a solution that has at least one trail that is not hamiltonian (say, T_1). By definition, there is at least one node (say, i) that does not appear in T_1. If there are multiple trails that are not hamiltonian, or if there are multiple nodes missing in a trail that is not hamiltonian, the argument that follows still holds good. Each of the $m - 1$ nodes have to appear before i at least once in the solution since each of these nodes have unit request to be sent to i. For each such node occurrence in a trail, the link that is incident from this node on the corresponding trail contributes a unit cost to the solution. Similarly, each of the $m - 1$ nodes have to appear after i in at least once in the solution and each link that is incident on this node will contribute a cost of 1 to the solution. The total cost of the solution is hence at least $(m-1) + (m-1) + 1$ (since T_1 costs at least 1 unit) $= 2m - 1$. This cost is strictly greater than the cost of the solution that has only trails that are hamiltonian which costs exactly $2m - 2$ as seen from Lemma 1.

3.3.2.2 Minimize Transceiver Cost Formulation

Define the decision version of the optimization problem as, given, $TRP_{tc} = <G, R, f, n + k>$, identify if there is a trail assignment in G, to satisfy R, with a cost of $n + k$ using cost function f. $TRP_{tc} \in NP$, since, for a certificate specifying the set of trails satisfying R, we can check in polynomial time, if the total number of transmitters and receivers active in the network is $n + k$ and if it is a legitimate assignment.

To prove that TRP_{tc} is NP-Hard, we describe the problem called $BIN - PACKING \in NPC$, and by using a reduction technique show that $BIN - PACKING \leq_p TRP_{tc}$. Suppose that there are n objects, where the size s_i of the ith object satisfies $0 < s_i < 1 : i \in (1..n)$. All the objects need to be packed into minimum number of unit sized bins. Each bin can hold any subset of the objects subject to the constraint that the total size of the objects does not exceed the bin capacity. $BIN - PACKING$ is a decision version of the problem which decides whether or not a given set of n objects, each of size, s_i can be packed into k bins. This problem is proved to be NP-Complete in [34].

The construction of graph G for the reduction process is as follows. For every object i, introduce a node v_i in the graph G. Introduce a sink node in the graph v_{n+1}. Define R such that the entries from v_i to v_{n+1} are marked $s_i * C$ while the rest are marked zero, where C is the capacity of the wavelength. Introduce an edge $(v_i, v_{i+1}) \forall i = 1..n$. Thus, the graph G is a simple path from v_1 to v_{n+1}.

Suppose that the given objects can be packed into k bins. Let A_i be the set of indices of the objects packed in bin i. Let min_i refer to the smallest element in A_i. For each bin i, establish a trail t_i from v_{min_i} to v_{n+1}. The routing is trivial since there exists only one path in the constructed topology. On trail t_i, requests of all objects in A_i can be served. This is possible since, the trail we just introduced allows all the nodes in A_i to transmit to v_{n+1} and the sum of request sizes does not exceed C. Since, there are k such bins, k circuits are required and each circuit is assigned a distinct wavelength. k receivers are activated on node v_{n+1}. A transmitter is required on each node $i, \forall i = 1..n$. Hence, the total cost of the network is $n + k$.

Suppose, the object cannot be packed in k bins, we claim that a network cost should be greater than $n+k$. This can be proved by contradiction. Suppose, that the cost of the network is $r(r \leq n+k)$. Since, each of the nodes, v_i through v_n have data to transmit to v_{n+1}, there are at least n transmitters, and hence the number of receivers (and consequently the number of trails) required to carry all the requests is at most k. For each trail t_i, let B_i denote the indices of the transmitters active on this trail. Each bin i can be assigned objects from B_i. This is possible since each bin can accept any subset of the objects, the sum of the requests in B_i did not overwhelm the capacity of trail t_i and each assignment is assigned to exactly one trail. So, the objects can be packed in at most k bins which is a contradiction to our assumption. Hence, if a trail assignment cost of $n+k$ is possible, the bin packing problem can be solved with k bins. This completes the reduction.

3.3.3 TRAW heuristics

Since TRP is NP-Hard, simple heuristics that yield fast and approximate solutions are required to design large scale optical networks. We propose three such heuristics and the general idea behind the heuristics is outlined below.

1. Select a set of candidate trails

2. Pack each candidate trail and assign it the required network resources

3. Choose the best candidate trail and place it in Γ

4. Update network status and repeat the above three steps until R is completely satisfied

The objective is to pack as many connections as possible onto a single trail to improve wavelength utilization. The output is the list Γ that contains the chosen trails and their assigned resources. The first step selects the set of candidate trails. The second step packs each candidate trail and assigns it the required resources. The three heuristics that we suggest differ primarily in the candidate selection criteria and in the wavelength assignment strategy.

The packing mechanism mentioned in the second step is the same for all heuristics. Recall that each connection is at most the size of wavelength capacity and there could be arbitrary number of connections between a source and a destination. Given a trail t, the containment set LT_t^r is listed out first in a sorted manner. The sorting could be done either in the increasing order or in the decreasing order but we observe that increasing order yields better results on an average. Select all elements in LT_t^r sequentially until choosing one more element violates the capacity of a wavelength. The selected connections can be packed on to the trail and the packing fraction for this trail, which is defined as the ratio of the used capacity to the trail capacity (wavelength capacity) is calculated. The range of the trail, that refers to the number of connections carried by this trail, is computed. A wavelength on every link and a transceiver on every node of the route is assigned to this trail on a first fit basis except in the third heuristic which uses a layered approach as will be explained later. Each of the candidate trails are packed and assigned resources in a way independent of the other candidate trails.

The third step chooses the best of the candidates and adds it to the list Γ. Choice of the best candidate is made keeping in mind the limited availability of transceivers and wavelengths in the network. For every candidate trail, the highest index of the wavelength (I_w) and the transceiver (I_t) allocated in the route are determined. The higher the index for a network resource, the more heavily the resource is used (typically) since we use the first-fit policy. The parameters taken into account for deciding the best candidates include I_w, I_t, range and packing fraction. Thus, every time, the trail that is likely to correspond to minimum congestion is chosen.

In the fourth step, traffic matrix R and used network resources are updated. These resources and requests will not be considered any further while packing candidate trails in the next iteration.

The above steps are repeated until each connection has been assigned to a trail. For all the heuristics below, a list L is required as input. Flloyd Warshall's algorithm is run and shortest routes between all node pairs with nonzero requests are computed. The node pairs are listed in the decreasing order of their shortest route lengths in list L. The specifics of the three heuristics are detailed below.

3.3.3.1 Longest of the Shortest routes first heuristic (LS)

Let (s,d) be first element in list L. It has the longest shortest route of length δ. There could be multiple routes between (s,d) and each of these routes could be considered to be a candidate trail. However, for this work, we only consider all possible routes from s to d of length δ to be candidates. Longer routes typically have bigger containment sets and hence may be able to accommodate more requests. Hence, the longest shortest paths are considered first.

Each candidate is packed and a first fit resource allocation is done. The best candidate is chosen with favorable I_w, I_t, packing fraction and range, in that order. The traffic matrix, the network resources and the load of the link are updated. The load of a link (i,j) is defined as $L_{i,j} = c(i,j) + \eta \ w_{i,j}$ where $c(i,j)$ is the cost of the original link (i,j), $w_{i,j}$ is the number of trails traversing link (i,j) currently and η is a parameter that can be tuned to achieve good performance. The candidate trails for next node pair in list L for the next iteration are identified by running shortest path algorithm considering current network load. By varying the load on the links dynamically for every iteration, congestion is minimized.

Let N be the number of nodes in the network. For each iteration, all pairs shortest path is run which is of complexity $O(N^3)$. Assume that only one element in list L is packed, it takes at worst $O(N^2)$ to locate that element. If there are a maximum of K requests from a node to another node, the size of the containment set of a LT is at most KN^2 and hence the packing heuristic, in the worst case, takes $O(KN^2 log(KN^2))$ time. The time complexity of this heuristic is $O(N^3 + KN^2 log(KN^2))$ for every iteration. with a maximum of $O(KN^2)$ iterations. If K is a constant, LS runs in $O(N^3)$ time per iteration.

3.3.3.2 Best Fit heuristic (BF)

The candidate trails correspond to all routes in L. Each of these trails are packed and assigned the necessary network resources. While choosing the best candidate, preference is given to the candidate with favorable I_w, I_t, packing fraction and range, in that order. The traffic matrix, network resources and link loads are updated as specified in LS heuristic. The current list L is removed and the routes between all node pairs are recomputed based on the current load and stored again in list L for the next iteration.

For each iteration, to find all pairs shortest paths using Flloyd Warshall's algorithm, it takes $O(N^3)$ time. Additionally, each item in list L has to be packed with the maximum size of the list being $O(N^2)$, thereby requiring a worst case time of $O(KN^4log(KN^2))$. Hence, the time complexity of BF is $O(KN^4log(KN^2))$ for every iteration with a maximum of $O(KN^2)$ iterations. If K is a constant, BF heuristic runs in $O(N^4log(N))$ time per iteration.

3.3.3.3 Layered Graph heuristic (LG)

In this heuristic, we view the WDM network as a multi layered network, one for each wavelength. We assume that each layer looks identical since resources are provisioned uniformly. There exists no wavelength conversion (no ladders) to hop from one layer to another. The first element (s,d) is considered and the candidate trails correspond to the routes between s and d on each layer. Packing of each trail is done as in previous two heuristics. Preference is given to I_t, packing fraction, and range, in that order. Once a trail is chosen on a particular wavelength, the links corresponding to that route are removed from that layer. The next node pair in L is considered for next iteration.

For each iteration, the shortest path heuristic and packing heuristic are run on all W layers. The shortest path heuristic costs $O((N^2logN)W)$ time and the packing heuristic costs $O(KWN^2log(KN^2))$. The complexity of the packing heuristic dominates and hence $O(KWN^2log(KN^2))$ is the worst case complexity of LG heuristic for every iteration with a maximum of $O(KN^2)$ iterations. If K is a constant, LG heuristic runs in $O(WN^2logN)$ per iteration.

The candidate set for BF heuristic is relatively large and hence has higher time complexity when compared to the other two. The LG heuristic maintains a network for every wavelength and has relatively higher space complexity. The LS heuristic is the simplest both in terms of space and time and will be fast and efficient for large scale networks.

3.4 Two-Hop Static Network Design

The resource sharing in light-trails is achieved as a result of drop and continue functionality and an overlaid control protocol. However, the new hardware introduces additional losses to the signal. The loss could be as high as 8 dB per node since the splitter and coupler suffers 3 dB loss while the shutter suffers a 2 dB loss. With lossy optical components, the signals may not always be carried end to end in the pure optical form. Even if compensation on every node is provided using amplifiers, due to amplifier noise, the signal may be degraded and not be able to traverse beyond a certain hop limit which we call the trail length limit (δ). In this case, it becomes necessary for a request to traverse multiple intermediate trails to reach the final destination.

Let $d_{i,j}$ and $R_{i,j}$ denote the distance and traffic between nodes i and j respectively. If $d_{i,j} > \delta$, pair (i,j) is said to be *physically blocked* and $R_{i,j}$ cannot be carried by a direct trail from i to j. There may exist an intermediate node k, such that $d_{i,k} \leq \delta, d_{k,j} \leq \delta$, in which case, k is said to be in the *proximity* of (i,j). If $R_{i,j}$ is first carried from node i to node k on one trail and then shifted to another trail from node k to node j, then we define k to be the *hub* for pair (i,j). In a static scenario, this multi-hop model is equivalent to a *traffic matrix rearrangement*, where the original traffic matrix is modified to obtain a new traffic matrix. That is, $R_{i,k}^n = R_{i,k}^o + R_{i,j}^o, R_{k,j}^n = R_{k,j}^o + R_{i,j}^o$ and $R_{i,j}^n = 0$, where the superscripts o and n refer to the old and new values respectively. The hub node (H-node) is just another node in the network, but equipped with special grooming hardware required to act as transit point for the physically blocked traffic.

The work in [39] allows every physically blocked pair (i,j) to choose a random node k in its proximity as its hub and hence requires all nodes to be grooming capable leading to significant network costs. In our work, we carefully design the H-node placement so that the total number of H-nodes is minimized while still not compromising on network throughput. It is important to emphasize here that while e-grooming in lightpaths is required to improve bandwidth utilization, such a functionality is already offered at the optical layer in light-trails. The hub nodes provide transit points for multi-hop traffic and we focus on minimizing such nodes.

3.4.1 Problem Definition

We give a formal description of the sparse hubbing problem here. Given a network topology
G(V,E), where V is the node set, E is the link set, C is the capacity of each wavelength, and
given the traffic matrix R, where each connection represents a sub-wavelength traffic, design a
network so as to optimize one of the following objectives: (1) For a given number of H-nodes
and wavelengths, maximize the network throughput. (2) Carry all the traffic while minimizing
the number of wavelengths and H-nodes used.

3.4.2 ILP Formulation

We formulate an integer linear program to solve the problem. We make the following assumptions in our study.There is no wavelength conversion capability. There exists at most one
fiber link between any node pair. Individual connection requests do not exceed the wavelength
capacity though aggregate traffic between a node pair can be of arbitrary value. Tunable
transmitters and wide bandwidth receivers are assumed to be present on all the nodes. A
connection request should never be split across multiple routes both on the physical and on
the logical topology. Each connection C_n, $1 \leq n \leq K$, is an ordered pair (s,d,p,y) which refers
to the p-th connection of granularity y from node s to node d. We define C_λ to be the cost of
maintaining a wavelength in the network and C_h to be the cost of a hub node. We describe
the rest of our notation below.

N - number of nodes in the network (data)

C - capacity of a wavelength (data)

W - number of wavelengths on each link of capacity C (data)

LT - set of possible light-trails in the network (data)

LT_t - an instance of a light-trail $LT_t \in LT$ (data)

LT_t^r - set of requests that can be supported by LT_t based only on the containment constraint
(data)

$LT_t^{i,j}$ - 1 if trail LT_t traverses link (i,j), 0 otherwise (data)

R_{c_n} - traffic request value of connection c_n (data)

$t = 1..\|LT\|$ - number assigned to each light-trail(index)

$\lambda = 1..W$ - number assigned to each wavelength (index)

$m, n = 1..K$ - number of connections in the network (index)

$i, j, s, d, = 1..N$ - nodes in the network (index)

α - a very large number (say, 10000) (data)

$\phi_{i,j}^{c_n}$ - 1 if c_n is carried by node pair (i,j), 0 otherwise (variable)

Φ^{c_n} - 1 if c_n is carried by the network, 0 otherwise (variable)

T_t^{λ} - 1 if wavelength λ is assigned to trail t, 0 otherwise (variable)

TX_t^i - 1 if node i on trail t needs a transmitter, 0 otherwise (variable)

RX_t^i - 1 if node i on trail t needs a receiver, 0 otherwise (variable)

$X_{i,j,t}^{c_n}$ - 1 if node pair (i,j) carries c_n on trail t, 0 otherwise (variable)

χ_i - 1 if a node is hub capable, 0 otherwise (variable)

U_{λ} - 1 if wavelength λ is used, 0 otherwise (variable)

N_h - number of hub nodes (variable)

N_{λ} - number of wavelengths used in the network (variable)

T_t - number of instance of trail LT_t (variable)

3.4.2.1 Maximize Throughput

$$\text{Maximize} \sum_{c_n} \Phi^{c_n} \tag{3.11}$$

Subject to the following constraints

Virtual Topology Constraints:

$$\sum_j^{j \neq s} \phi_{s,j}^{c_n} = \Phi^{c_n} \quad \forall c_n \tag{3.12}$$

$$\sum_j^{j \neq d} \phi_{j,d}^{c_n} = \Phi^{c_n} \quad \forall c_n \tag{3.13}$$

$$\sum_{j}^{j \neq s,d} \phi_{j,i}^{c_n} = \sum_{j,j \neq s}^{i \neq s,d} \phi_{i,j}^{c_n} \quad \forall c_n, \forall i \tag{3.14}$$

$$\sum_{j} \phi_{j,s}^{c_n} = 0 \quad \forall c_n \tag{3.15}$$

$$\sum_{j} \phi_{d,j}^{c_n} = 0 \quad \forall c_n \tag{3.16}$$

$$\chi_i \geq \sum_{j} \sum_{c_n}^{j \neq s} \phi_{i,j}^{c_n}/\alpha \quad \forall i, \ i \neq s \tag{3.17}$$

$$N_h \geq \sum_{i} \chi_i \tag{3.18}$$

Physical Route Assignment Constraints:

$$\sum_{t} X_{i,j,t}^{c_n} = \phi_{i,j}^{c_n} \quad \forall c_n, \forall i, (i,j) \in LT_t^r \tag{3.19}$$

$$\sum_{i,j \in LT_t^r} \sum_{c_n} R^{c_n} X_{i,j,t}^{c_n} \ \leq \ T_t \, C \quad \forall t \tag{3.20}$$

Wavelength Assignment Constraints:

$$\sum_{\lambda} T_t^{\lambda} = T_t \quad \forall t \tag{3.21}$$

$$\sum_{t} T_t^{\lambda} \leq 1 \quad \forall \lambda, \{t : LT_t^{p,q} = 1, \forall (p,q) \in E\} \tag{3.22}$$

$$U_{\lambda} \geq \sum_{t} T_t^{\lambda}/\alpha \quad \forall \lambda \tag{3.23}$$

$$N_{\lambda} \geq \lambda \, U_{\lambda} \quad \forall \lambda \tag{3.24}$$

Transceiver Usage Constraints:

$$TX_t^i \geq X_{i,j,t}^{c_n} \quad \forall c_n, \forall t, \forall i, \forall (i,j) \in LT_t^r \tag{3.25}$$

$$\sum_t TX_t^i \leq TX_i \quad \forall i \in LT_t \tag{3.26}$$

$$RX_t^i \geq X_{j,i,t}^{c_n} \quad \forall c_n, \forall t, \forall i, \forall (j,i) \in LT_t^r \tag{3.27}$$

$$\sum_t RX_t^i \leq RX_i \quad \forall i \in LT_t \tag{3.28}$$

Traffic non-bifurcation Constraints

$$Y_t^{c_n} \geq \sum_{(i,j) \in LT_t^r} \lambda_{i,j,t}^{c_n}/\alpha \quad \forall c_n, t \tag{3.29}$$

$$Y_t^{c_n} \leq \sum_{(i,j) \in LT_t^r} \lambda_{i,j,t}^{c_n} \quad \forall c_n, t \tag{3.30}$$

$$I_t^{c_n,c_m} \leq (Y_t^{c_n} + Y_t^{c_m})/2 \quad \forall c_n, c_m, t \tag{3.31}$$

$$I_t^{c_n,c_n} = (Y_t^{c_n} + Y_t^{c_n})/2 \quad \forall c_n, t \tag{3.32}$$

$$R_{c_n} + \sum_{c_n, c_m \neq c_n} R_{c_m} I_t^{c_m,c_n} \leq C \quad \forall c_n \tag{3.33}$$

$$S_t^{c_n} \leq \sum_{m=1}^{n-1} I_t^{c_n,c_m} \quad \forall c_m, t \tag{3.34}$$

$$S_t^{c_n} \geq \sum_{m=1}^{n-1} I_t^{c_n,c_m}/\alpha \quad \forall c_m, t \tag{3.35}$$

$$T_t = I_t^{c_1,c_1} + \sum_{n=2}^{K}(I_t^{c_n,c_n} - S_t^{c_n}) \quad \forall t \tag{3.36}$$

$$X_{i,j,t}^{c_n}, \phi_{i,j,t}^{c_n}, \chi_i, U_\lambda, \Phi^{c_n}, T_t^\lambda \in (0,1) \qquad\qquad TX_t^i, RX_t^i, N_h, N_\lambda, T_t \in I \qquad\qquad (3.37)$$

The above formulation accepts set of possible trails (LT) in the network as input and maximizes throughput. Equation (3.11) maximizes the carried connection. Equation (3.12), (3.13), (3.14), (3.15) and (3.16) route all the accepted connections using the flow conservation on the source, destination and any intermediate node. Equation (3.17) identifies the hub nodes and equation (3.18) counts the total number of hub nodes in the network. Equation (3.19) determines the physical route (specified by the trail) on which the connections are carried. Equation (3.20) allows a wavelength to be packed only up to its maximum capacity and equation (3.21) assigns a wavelength to each trail. Equation (3.22) prevents wavelength collision between trails sharing a link while equation (3.23) keeps track of the wavelengths that are used in the network and equation (3.24) counts the total number of wavelengths required. A node may have a trail traverse it but may still not be active on it since the signal bypasses the LAU on that node as discussed in section 2.1.2.7. Equations (3.25) and (3.27) identify the trails on which the node i are active for transmission and reception respectively. Equations (3.26) and (3.28) ensure that the number of communication equipments required do not exceed the resources provisioned.

The above constraints prevent a connection from being split on the virtual topology. However, it does not ensure that the connection is not bifurcated on the physical topology when there are multiple instances of the same trail. To understand this, consider a two node network where i and j are the nodes and (i,j) is the only link. Assume three connections (c_1,c_2,c_3) from i to j, 6 units each, and a wavelength capacity of 10 units. Based on equation (3.20), the aggregate capacity required is 18 units and hence can be carried on two trails. However, it is clear that three trails are required if the connections must not to split and this is not captured in any of the constraints above. This problem was first observed and addressed in the context of in [66] and we adopt a similar approach here .

Equation (3.33) ensures that the sum of traffic carried by any instance of a trail never exceeds the capacity of the wavelength and hence prevents bifurcation. For instance, in the above example, if any two connections, say, c_1 and c_2 happen to share the same trail, then the entry $I_t^{c_1,c_2}$ is one and hence will violate the capacity constraint (3.33). The final constraint that relates the non-bifurcation aspect to the physical routes is (3.36). The right side of the constraint counts the number of instances of the same trail required between two nodes to carry the connections just by looking at the I matrix and avoids multiple counting by introducing a variable $S_t^{c_n}$. A more detailed description can be found in [66].

3.4.2.2 Minimize Cost

If the objective function in the above formulation is replaced with the new function **Minimize** $C_\lambda N_\lambda + C_h N_h$, and if $\Phi^{c_n} = 1 \ \forall c_n$, is set as an additional constraint, while retaining the rest, the new formulation optimizes cost of a sparsely hubbed network, while accepting all the traffic. If the trail length is constrained, only trails of restricted length may be provided as input (LT) to the ILP. All the physically blocked requests are automatically and optimally assigned hubs by the ILP. If N_h is removed from the objective function, it reflects light-trails with full hubbing capability and now if $N_h = 0$ is introduced as an additional constraint, it models light-trails with no hubbing capabilities.

3.4.3 Heuristics for Two-Hop Sparse E-grooming

The ILP problem is computationally intractable and is not feasible for design of large networks. We propose heuristics to solve the sparse grooming problem in such cases. For a given network topology, number of wavelengths and H-nodes, our heuristics work as follows. We first choose the nodes to be equipped with hubbing capabilities based on some selection criteria. We perform traffic matrix rearrangement and carry physically blocked connections through hub nodes (called hubbing). The connections are then routed on the physical topology using a trail routing heuristic. Finally, a first-fit wavelength assignment is done for each trail. We outline only the main steps of the heuristics due to space constraints.

3.4.3.1   H-node Selection

We suggest three criteria to select H-nodes in the network.

1. Eccentricity criteria (EC): The eccentricity of a vertex in a graph is the longest of the shortest paths between the vertex and all the other points in the network. Nodes with low eccentricity values can be good H-node candidates.

2. Proximity criteria (PC): Find the number of physically blocked (i,j) pairs such that a node k lies in the proximity of (i,j) and assign this as a rating for k. Nodes with high proximity rating can be good H-node candidates.

3. Random criteria (RC): H-nodes are randomly chosen.

3.4.3.2   Hubbing

The physically blocked node pairs are sorted in list B in the non-increasing order of their aggregate request values. The set of hub capable nodes (selected based on one of the above mentioned criteria) are ordered in list H. Consider the first node pair (i,j) ∈ B. Find all candidates k ∈ H such that k lies in the proximity of (i,j) and there already exists some traffic from i to k and k to j. Make an entry for k in list S. The blocked node pair (i,j) is then hubbed by one or many of the candidate nodes in S since individual connections from (i,j) can traverse different paths to reach the destination. The hubbing heuristic rearranges traffic and tries to accommodate it without creating the necessity to open up a new trail unless absolutely required. This procedure is repeated for every (i,j) in list B until no further traffic rearrangement is possible. If there still exists some physically blocked pairs that are not rearranged, they remain blocked. A node pair (i,j) could be blocked because there may not exist any node in list H that lies in the proximity of this node pair. The rest of the rearranged requests are packed onto trails and routed subject to the wavelength availability constraint.

3.4.3.3   Trail Routing and Wavelength Assignment

The trail routing and wavelength assignment heuristic used here is similar to the LS heuristic except that we deal with connections of multiple granularities subject to non-bifurcation constraints. The prime focus of the heuristic is to pack as many requests as possible onto a trail while still balancing the load on the links. It runs the Flloyd Warshall's algorithm and finds the shortest path between all possible node pairs. The hubbing step ensures that most of the physically blocked pairs with non-zero traffic have their connections rearranged through the hub nodes.

Based on the rearranged matrix, sort all the node pairs that are not physically blocked in the non-increasing order of their shortest path lengths in list L. The farthest node pairs in L are routed first. Multiple shortest paths are tried and a first-fit wavelength assignment is performed selecting the route that corresponds to the lowest wavelength index, since this minimizes congestion. In case of a tie, the trail that is maximally packed (described below) is chosen. The traffic matrix is updated to reflet the routed connections. The node pairs are scanned sequentially in L, and the process of identifying the next trail is repeated in a similar manner until no more trails can be routed because of wavelength exhaustion.

A chosen trail LT_t is packed in the following way. Recall that each node pair has many requests of sub-wavelength granularities. The aggregate traffic between this node pair may have any arbitrary value. The fraction of the requests between the start node and end node of the trail, which can be accommodated on one wavelength, while adhering to the capacity and non-bifurcation constraints, are first packed onto this trail. The other elements belonging to the containment set of the trail t are listed in the non-decreasing order of their aggregate requests in list LT_t^r. All the items from the left in list LT_t^r are selected sequentially until taking one more item (p) may exceed the capacity of trail LT_t. The trail may still have some residual capacity. Consider the last node pair p whose aggregate traffic is too big to be accommodated on this trail. This node pair, p, may have some individual connections that are small enough to be accommodated on LT_t. Sort these individual connections in the non decreasing order and pack as many as possible on LT_t without violating the capacity constraint. Once, no additional connection of this node pair p can be accommodated on LT_t, check if the trail still has residual capacity. Proceed in a similar way with the next node pair q by sorting the individual connections of node pair q and repeat the above step for every subsequent node pair in LT_t^r until no more connections in the set can be carried on this trail.

3.5   Multi-hop Dynamic Network Dimensioning

In this section, we consider dynamic traffic scenarios wherein call arrivals and call holding times follow a well known stochastic distribution. The objective of any RWA algorithm in such a scenario is typically to maximize the acceptance likelihood for an incoming call. We try to minimize blocking probability for future requests by routing the current requests in the most economical way that is possible taking into account the cost of the network resources. An algorithm that is designed to solve this problem should account for the various levels of heterogeneities that may be found in WDM networks. Heterogeneities in networks may arise [150] since equipment from different vendors may be used or new equipment may need to coexist with old equipment or network upgrades may result in only selected parts of the network being modified. For instance, the following variations in WDM network could be found:

- Different nodes may be equipped with different number of transceivers

- Different links may be equipped with different number of wavelengths

- Different OXCs may have different grooming capabilities

- Wavelength conversion and grooming capabilities may be sparsely available in the network.

An algorithm that deals with network design should be easily extensible to model all these different scenarios. In this section, we develop a unified graph based heuristic that can model all these heterogeneities and yield results for any optical network that employs path level aggregation strategy.

The work in [138] also uses an auxiliary graph approach to solve the multi-hop design problem in LT networks and is the most relevant research to our current work. In both the graph based approaches, for each physical node in the network, a number of vertices are introduced in the auxiliary graph to model the state of the network. In [138], the number

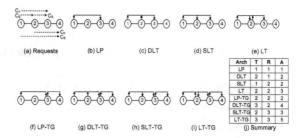

Figure 3.4   (a) Request sequence. (b) Only C_1 is accepted. (c) Only C_1 and C_3 are accepted. (d) Only C_1 and C_2 are accepted. (e) Only C_1, C_2 and C_3 are accepted. (f) Only C_1 and C_5 are accepted. (g) Only C_1, C_3, C_4 and C_5 are accepted. (h) Only C_1, C_2 and C_5 are accepted. (i) all connections are accepted. (j) summary: T - number of used transmitters, R - number of used receivers, A - number of accepted connections

of vertices introduced in every wavelength layer is proportional to the number of physical links while in our model, it is proportional only to the number of physical nodes. Our model is applicable for any path level aggregation strategy unlike the earlier work which models only light-trails. The authors of [138] focused only on the service provisioning aspects. The contribution of this work is to introduce many different metrics and compare the performance of various grooming and non-grooming PLATOON architectures for these various metrics based on our auxiliary graph model. With our proposed model, we are able to identify scenarios for which one architecture would outperform the other, while such conclusions were not possible in the earlier work.

The rest of this section is organized as follows. A five node sample network scenario in which a sequence of calls need to be routed are provided in Section 3.5.1 and the resource requirements of architectures with and without electronic grooming capabilities are analyzed. An electronic grooming switch design for LT architectures is presented in Section 3.5.2. Finally, Section 3.5.4.2 develops the auxiliary graph based model and works out a small three node network example to illustrate the grooming heuristic.

3.5.1   Resource Consumption in PLATOONs

The network dimensioning problem in the context of dynamic traffic can be defined as follows. Given the physical topology, a fixed number of wavelengths per link (W) and a fixed number of transceivers per node (X), identify the most economical route and wavelength assignment of an incoming request so as to maximize the probability of accepting future requests. If the client layer is equipped with electronic traffic grooming (TG) capabilities, connections may go through several OEO conversions (and hence multiple hops) before reaching the final destination. If TG capabilities are absent, connections reach the destination with a single hop. To refer to an architecture with TG capability in the client layer, we suffix the architecture name with the letters TG. In our work, using a unified framework, we solve the dimensioning problem for eight architectures - LP, SLT, DLT and LT and their respective counterparts that are capable of full electronic grooming - LP-TG, SLT-TG, DLT-TG and LT-TG and compare their performance.

The performance of these architectures for an example four node network with one wavelength per link and 3 transceivers per node is shown in Figure 3.4. Let C_i refer to the ith call in a sequence of five call arrivals. Each call has a bandwidth requirement of 2 units and the capacity of the wavelength is 10 units. The resource requirements for all the networks are shown in Figure 3.4.

This example brings to light the trade-offs involved in o-grooming. Though o-grooming increases aggregation capability, it comes with a price which we call aggregation penalty. There are two kinds of aggregation penalties - bandwidth penalty and transceiver penalty. Due to these penalties, even if packing of requests into circuits are done in the ideal way, transceiver or wavelength utilization of 100 % is not achievable if there is more than one source or sink in a circuit. For instance, consider the DLT-TG circuit (say, T) in Figure 3.4(g), that carries two connections - 1 to 3 and 2 to 3. A total of three communication units are used - a receiver on node 3, a trasmitter each on nodes 1 and 2. The wavelength is capable of filling only a maximum of 10 units of transmitter capacity and 10 units of receiver capacity. By tuning a second transmitter to accept C2, the circuit has a total of 20 units of transmitter

94

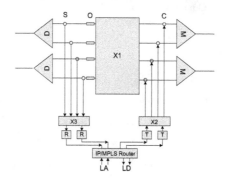

Figure 3.5   Partial grooming architecture for light-trail networks. D - De-
multiplexer, M - Multiplexer, O - Optical Shutter, X1,X2,X3
- OXC, T - Transmitter, R - Receiver, LA - Local Add, LD -
Local Drop, S - Splitter, C - Combiner

capacity, of which 10 units will remain unusable. The network has a total transceiver capacity
of 240 units (3 transmitters and 3 receivers for each of the four nodes), of which 4.16 % (10
units) is unusable, and this is called the transceiver penalty (scaled by the network transceiver
capacity). The bandwidth penalty arises because C3 locks up 2 units of bandwidth on the
entire wavelength, that is, including link (1,2). This bandwidth locking beyond the connection
span is required since the circuit, by intentional design, does not support optical packet level
switching. The total bandwidth consumed on all the links of the network is 18 units (6 units
for C_5 and 4 units each for C_1, C_3, and C_4). The bandwidth penalty (scaled by the total
consumed bandwidth) is 11.11%. The bandwidth and transceiver penalties are not incurred
for LP networks since they have exactly one source and sink.

3.5.2   E-Grooming Switch Architecture for PLATOONs

The e-grooming architecture in the context of light-trail networks is discussed in this section.
Figure 3.5 shows a grooming switch architecture which allows for up to 4 wavelengths to be
shared. It allows the transmitters and receivers to be used independently. Consider a node
that is active on four trails (T_1, T_2, T_3 and T_4), where it transmits only on T_1 and T_2 and

receives only on T_3 and T_4. Assume that traffic from T_4 needs to be groomed onto T_1 and that none of the trails start or end at this node. The four trails are tapped at the local node, but T_3 and T_4 are selected by X3 to be detected at the two receivers. All the four trails continue through their LAUs. Signals from the two local transmitters are switched by X2 to trails T_1 and T_2. The IP/MPLS router grooms connections across trails through a software based queueing scheme. Please note that a MAC unit is required to access the channel (not shown in the figure). This grooming architecture can be extended to model SLT and DLT networks as well and we do not present it here. E-grooming architectures for lightpath networks can be found in [146].

3.5.3 Problem Definition

The network design in the context of dynamic traffic can be formally defined as follows. The call arrivals and departures at the network can be modeled using well known stochastic distributions. Requests of varying bandwidth granularities arrive between different source destination pairs at a network. The bandwidth requirement and the source-destination tuple for a request also follows a known stochastic distribution. Given the physical topology, a fixed number of wavelengths per link and a fixed number of transceivers per node, identify the route and wavelength assignment of an incoming request so as to maximize the probability of accepting future requests.

3.5.4 The Auxiliary Graph Model for PLATOONs

In this section, we first introduce the auxiliary graph approach and outline the various steps of the algorithm. Next, we provide the rules based on which the auxiliary graph is generated and describe the virtual topology representation for various network architectures. We develop the algorithm for LT-TG (referred to as trails below) though it will be shown that with minor modifications in the virtual topology representation, other architectures can be modeled as well. Subsequently, we identify the grooming policy and assign costs to network resources to achieve the required objectives. Finally, we illustrate the heuristic with an example six node

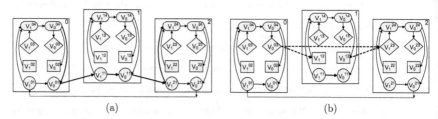

Figure 3.6   (a) Three node single wavelength unidirectional LT-TG ring (b)
Connection from node 0 to node 2 is set up on trail {0,1,2}

network.

3.5.4.1   The Auxiliary Graph Approach

The basic idea behind our model is as follows. We design an algorithm that takes traffic request $T(s,d,m,t_a,t_e)$ as input, where m is the value of the subwavelength request between the source s and destination d, t_a is the arrival time and t_e is the exit time of the call. An auxiliary graph which is also alternatively known as the reachability graph is obtained based on the current network state taking into account the availability of grooming switches, transceivers, wavelengths and their respective costs. The cost of resources are fixed based on some chosen grooming policy. Using a simple Dijkstra's algorithm, the shortest route is identified between s and d that can accommodate the request m. The concatenation of links in the shortest route constitute the physical route and the wavelength assignment for the request. If such a route does not exist, the required resources in terms of wavelengths or transceivers or switching ports may not available in the network to carry the call, and hence the call may be blocked. Otherwise, the network state is modified to reflect the resources consumed by admitting the current request and the algorithm proceeds to handle the next traffic request.

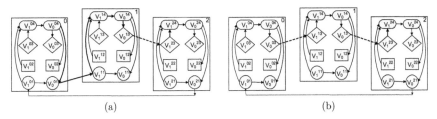

(a)   (b)

Figure 3.7   (a) Connection from node 0 to node 2 is torn down. Connection from node 1 to node 2 remains. (b) New connection from node 0 to node 2 is admitted. New trail {0,1} is set up and connection groomed through node 1

3.5.4.2   The Auxiliary Graph Generation

The physical topology of a network can be represented by a graph $G'(V', E')$ where V' is the node set and E' is the link set. The auxiliary graph corresponding to the physical graph is defined as $G(V, E)$, where the $|V|$ is the number of vertices introduced to represent the nodeset and $|E|$ is the number of edges introduced to describe the network state.

Node Model

In G, each node comprises of $W + 3$ layers with each layer including an input port and an output port. Layers 1 through W are the Wavelength layers (WLs), layer $W + 1$ and layer $W + 2$ are the Trail layers (TLs) or the Virtual Topology layer (VTL), and layer $W + 3$ is the Grooming layer (GL) or the Access Layer (AL).

Network Model

G is a graph with $(2W + 6)V'$ vertices and is generated as follows. Let $V_y^{i,k}$ refer to the yth port on layer k at node i. Let $y = 1$ refer to the input port and $y = 0$ refer to the output port. Eight different types of edges are inserted in the auxiliary graph based on the network state.

- E-Grooming edge: If the network has e-grooming capabilities, for each node i, there is an edge introduced from the input port to the output port of the Grooming Layer. That is,

$$(V_1^{i,W+3}, V_0^{i,W+3}) \in E \quad \forall i \in V' \tag{3.38}$$

For modeling networks with no e-grooming capabilities, the e-grooming edge is not introduced.

- Wavelength link edge: If a free wavelength k exists on the link (i, j), an edge is introduced between the output port of wavelength layer k on node i and the input port of wavelength layer k on node j. That is,

$$(V_0^{i,k}, V_1^{j,k}) \in E \quad \forall (i,j) \in G',$$

$$wavelength \ k \ is \ free \ on \ (i,j) \tag{3.39}$$

The capacity of this edge is the capacity of a wavelength.

- Trail edge: Let z denote a communication unit on a node with $z = 0$ referring to a transmitter and $z = 1$ referring to a receiver. Let T_t refer to the tth trail in the network. If $i \in T_t$, node i is active on the trail. Define $A_t^z(i) = 1$ if $i \in T_t$ and node i's z^{th} communication unit is active on T_t; 0 if $i \in T_t$ and node i's z^{th} communication unit is idle on T_t; -1 otherwise. Let $I_t(j)$ refer to the location of node j in trail T_t. Four kinds of edges are introduced for LT networks based on the following conditions.

$$(V_0^{i,W+1}, V_1^{j,W+1}) \in E, for \ some \ trail \ t \ carried \ in \ G,$$

$$\forall i, j : A_t^0(i) = 0, A_t^1(j) = 0, I_t(j) > I_t(i) \tag{3.40}$$

$$(V_0^{i,W+1}, V_1^{j,W+2}) \in E, for \ some \ trail \ t \ carried \ in \ G,$$

$$\forall i, j : A_t^0(i) = 0, A_t^1(j) = 1, I_t(j) > I_t(i) \tag{3.41}$$

$$(V_0^{i,W+2}, V_1^{j,W+1}) \in E, for \ some \ trail \ t \ carried \ in \ G,$$

$$\forall i, j : A_t^0(i) = 1, A_t^1(j) = 0, I_t(j) > I_t(i) \tag{3.42}$$

$$(V_0^{i,W+2}, V_1^{j,W+2}) \in E, \ for \ some \ trail \ t \ carried \ in \ G,$$

$$\forall i, j : A_t^0(i) = 1, A_t^1(j) = 1, I_t(j) > I_t(i) \tag{3.43}$$

Equation (3.43) states that if i's transmitter unit is active on t, j's receiver unit is active on t and j is downstream of i on trail t, an edge is introduced from output port of layer $W+2$ on node i to input port of layer $W+2$ on node j. Similar interpretations can be extended for the other equations as well. The capacity on all these edges is the residual capacity of trail t. A more detailed description of trail edges will be illustrated using an example in a later section. The above mentioned equations are true only for LT and LT-TG networks.

For modeling other networks, the following constraints need to be added for equations (3.40) to (3.43). For modeling SLT and SLT-TG networks, include the additional constraint that i should be the convener node for trail t. For modeling DLT and DLT-TG networks, include the additional constraint that j should be the end node for trail t. For modeling LP and LP-TG networks, include the constraint that i should be the convener node and j should be the end node for trail t.

- Mux edge: There is an edge introduced between the output port of layer $W+3$ on node i to output port of layer $W+2$ on node i.

$$(V_0^{i,W+3}, V_0^{i,W+2}) \in E \quad \forall i \in V' \tag{3.44}$$

- Demux edge: There is an edge introduced between the input port of layer $W+2$ on node i to input port of layer $W+3$ on node i.

$$(V_1^{i,W+2}, V_1^{i,W+3}) \in E \quad \forall i \in V' \tag{3.45}$$

- Receiver edge: If there is at least one free receiver available at node i, two types of arcs are introduced. First, there is an edge introduced from the input port of every WL at node i to the input port of GL at node i. Second, an edge is introduced between input port of layer $W+1$ at node i to input port of GL at node i.

$$(V_1^{i,k}, V_1^{i,W+3}) \in E \quad \forall i \in V', k \in \{1,..W\} \tag{3.46}$$

$$(V_1^{i,W+1}, V_1^{i,W+3}) \in E \quad \forall i \in V' \tag{3.47}$$

- Transmitter edge: If there is at least one free transmitter available at node i, two types of arcs are introduced. First, there is an edge introduced between output port of GL at node i to output port of every WL at node i. Second, there is an edge introduced between output port of GL at node i to output port of layer $W + 1$ at node i.

$$(V_0^{i,W+3}, V_0^{i,k}) \in E \quad \forall i \in V', k \in 1,..W \tag{3.48}$$

$$(V_0^{i,W+3}, V_0^{i,W+1}) \in E \quad \forall i \in V' \tag{3.49}$$

- Converter edge: If wavelength k_1 is convertible to k_2 at node i, there is an edge between input port of layer k_1 and the output port of layer k_2 on node i.

$$(V_1^{i,k_1}, V_0^{i,k_2}) \in E \quad \forall i \in V'$$

$$\text{wavelength } k_1 \text{ is convertible to } k_2 \text{ on node } i \tag{3.50}$$

- Wavelength bypass edge: There is an edge from the input to the output port of each wavelength layer at node i

$$(V_1^{i,k}, V_0^{i,k}) \in E \quad \forall i \in V', k \in \{1..W\}$$

Each edge (i, j) in the auxiliary graph is associated with a tuple $\{c_{i,j}, w_{i,j}\}$, where $c_{i,j}$ refers to the residual capacity of the edge and $w_{i,j}$ refers to edge's weight or cost. The capacity of a wavelength edge is the capacity of a wavelength. For the trail edge, the capacity is the residual capacity of the trail. For all the other edges, the capacity is assigned to be infinity. The trail edges carry additional information pertaining its route and wavelength assignment. The cost of the edges can reflect the actual cost of a network element and/or the requirements of a grooming policy.

3.5.4.3 Virtual Topology Representation

The virtual topology captures the reachability information related to the established circuits in the network. Each node has two layers and both layers have a transmitter node and a receiver node. The layers are called *Busy* and *Idle* layers. A transmitter (receiver) unit in the *Busy*

STEP 1: Generate a request $T(s, d, m, t_a, t_e)$

STEP 2: Identify the list S of calls in the network that are scheduled to leave before t_a.

STEP 3: consider a call $s \in S$.

STEP 4: Remove the call from the network. Update the transceivers consumed by this call on all the nodes in its route and the wavelengths used by this call on all the links in its route. If the number of transmitter units available on a node just became nonzero, an edge needs to be introduced between the output port of GL on this node to the W input ports of the WLs on this node. An edge is also introduced between the output port of GL on this node to the output port of layer $W + 1$ on this node. Repeat the procedure with the receiver units modifying appropriate arcs on appropriate nodes.

STEP 5: Update all the trails that carried this call. Some of the corresponding trail edges in the auxiliary graph may not be consistent with equations (3.40) - (3.43) since the status of the transceivers on some nodes in the network may have changed. Move the origination or termination port of these edges so that they are consistent with the new state specified by the equations (3.40) - (3.43). Update the property tuple $(c_{i,j}, w_{i,j})$ for the remaining trail edges. Dimension or trim the trails and free some of the wavelength links, if possible.

STEP 6: Remove call s from list S. If S is empty proceed to STEP 7 else return to STEP 3.

STEP 7: Delete the edges in G whose residual capacity is less than m since they cannot carry this call.

STEP 8: Find the shortest route P from the output port of GL layer on node s to input port of GL layer on node d based on the grooming policy.

STEP 9: If P does not exist drop the call and return to STEP 1

STEP 10: If P exists, decompose P into two sets N and O, where N is the list of new trails to be created and O is the list of old trails to be updated to carry this call.

STEP 11: For $n \in N$, the wavelength link edges denoting the wavelength links used by the call are removed from the wavelength layers. Trail edges are introduced in G based on the equations (3.40) - (3.43). The cost of the new trail is defined by the grooming policy and the residual capacity of the trail edges are reduced by the value of the traffic routed on this trail. Update the transceiver requirements on all the nodes of the trail.

STEP 12: For every element $o \in O$, update the transceiver status on all the nodes of the trail. This may make some of the existing trail edges inconsistent with the equations (3.40) to (3.43). Shift the origination and termination points of these edges until they are consistent with the new state specified by equations (3.40) to (3.43). Reduce the residual capacity of all the trail edges corresponding to this trail by the amount of traffic routed.

STEP 13: If the number of free receivers on any node in the path P becomes zero, remove the receiver edge from the W input ports of the WLs on this node to the input port of GL on this node. Also, remove the receiver edge that goes from the input port of layer $W+1$ on this node to the input port of GL on this node. Repeat the procedure with the transmitter unit modifying appropriate arcs on appropriate nodes.

STEP 14: Return to STEP 1.

Figure 3.8   Auxiliary graph based traffic grooming heuristic

layer is already provisioned for a circuit. This means that there is at least one connection in the circuit that is being sourced (sunk) by this node. A transmitter (receiver) unit in the *Idle* layer is not provisioned yet but can be tuned into the circuit. The reason for having two layers is explained using the following example.

Consider a three node LT network shown in Figure 3.9(a) with no electronic grooming capabilities. Initially, there are no calls in the network and hence the VTL is devoid of any links. Consider a call that arrives between node 0 and node 1. The call is routed along the link 0-1 and circuit C1 is established. In the VTL shown in Figure 3.9(b) , the link (B_o^0, B_i^1) is introduced and the residual capacity of the circuit is calculated. Consider the next call that arrives between node 0 and node 2. The call is routed along the links 0-1-2 and is called circuit C2. The reachability information corresponding to the containment set of the circuit is captured in the VTL as shown in Figure 3.9(b). Since C2 is an LT circuit, arcs (B_o^0, I_i^1), (I_0^1, B_i^2) and (B_o^0, B_i^2) are introduced. If the circuit were SLT, only the first and the third arcs are introduced. If the circuit were DLT, only the second and third arcs are introduced. If it were simply an LP circuit, the third arc alone is introduced.

Consider a new call that arrives between node 0 and node 1. Assume that there is sufficient residual capacity in both the circuits. It is clear that both C1 and C2 can accommodate the call, since there are two arcs from node 0 to node 1. However, it may be a better idea to use C1 since the required transmitter and receiver units are already provisioned. It is important to note that in Figure 3.9(b), the busy transmitter unit on node 0 is connected to the idle receiver unit on node 1 and to the busy receiver unit on node 2. If a call from node 0 to node 1 is to be established on this circuit, the model representation makes it clear that a new receiver needs to be provisioned on node 1. In our model, we assign a large cost for an arc from idle receiver node to access layer and a small cost for an arc from busy receiver node to access layer. In setting up a call, an end to end route needs to be identified. This ensures that circuit C1 is chosen over circuit C2 in setting up this new call since the specified cost model makes the circuit C1 more favorable.

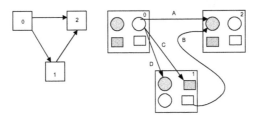

Figure 3.9   (a) Three node LT network (b) Virtual topology after connection from node 0 to node 2 is set up on trail {0,1,2}

3.5.4.4   The Auxiliary Graph Algorithm

The input to the reachability graph algorithm shown in Figure 3.8 is the current network state in the form of the auxiliary graph generated as mentioned above. The method takes a call $T(s, d, m, t_a, t_e)$, runs a shortest route algorithm on the auxiliary graph from the output port of GL on node s to the input port of GL on node d. The call can be routed on a network in many ways without connection rerouting. The route generated by Dijkstra's algorithm could be one of the following: (a) It is a direct trail from source to destination. (b) It is a concatenation of multiple trails from source to destination.(c) It is a concatenation of multiple free wavelength links from source to destination. (d) It is a concatenation of both trails and free wavelength links from source to destination. Note that the choices (b) and (d) arise only in the case with e-grooming capabilities. If a source to destination route exists, the call is accepted, the route and wavelength are assigned, the auxiliary graph is updated and the algorithm proceeds to handle the next request. Otherwise, the call is blocked.

If a call is to be torn down, the residual capacity of the corresponding trail is updated and the trail is dimensioned accordingly. If required, the trail is torn down and the corresponding wavelength links are freed. The running time of this set up algorithm is $O(|E|log|V|)$ since we use heap implementation of shortest paths.

3.5.4.5   Cost Model

This auxiliary graph model can achieve various objectives using different grooming policies. When there are multiple routes between the source and the destination, it is possible that we may prefer one of the routes over the other for carrying traffic and this reflects the chosen grooming policy. Typically, we specify a grooming policy for network operation in terms of the parameter we want to optimize and try to achieve it by prioritizing the route selection process based on this parameter and greedily applying it to every request carried by the network as explained in [145].

The cost of the links in the auxiliary graph is specified based on the grooming policy. The cost of the link could be static or dynamic. A static grooming policy may be to minimize the number of consumed transmitters or wavelengths. Such a policy can be implemented using the dominant edge principle. If, for instance, the objective is to minimize transceiver consumption, the unused transmitter and receiver edges are assigned a high cost while the other edges are assigned a unit cost. This will ensure that new transceivers are used only when absolutely required. An example dynamic policy would be to make the unused transmitter and receiver edges dominant while the cost of the wavelength edge passing through link i is given by,

$$WE_i = \eta T_i$$

where T_i is the number of circuits passing through link i and η is empirically chosen based on simulations. While the transceiver consumption is limited by the high cost of the transceiver edges, the wavelength link cost does load balancing and reduces wavelength consumption as well. We use the adaptive grooming policy in our algorithm.

3.5.4.6   Illustrative example

We describe the auxiliary graph generation procedure using a single wavelength three node unidirectional LT-TG ring {0,1,2} with the ring direction specified by the arc (0,1). All the connections that arrive have half the capacity of the wavelength and hence two of them can share one wavelength. Initially, no trail has been set up in the network. Assume that a connection $C_{0,2}^1$ arrives at the network. This request can be routed through the available free wavelength links on the route $T_1 = \{0,1,2\}$ shown in Figure 3.6(a). As a result, free wavelength links $(V_0^{0,1}, V_1^{1,1})$ and $(V_0^{1,1}, V_1^{2,1})$ are removed and a trail edge $(V_0^{0,3}, V_1^{2,3})$ is introduced in G.

It can be seen that the newly established trail can accommodate the request $(0,1)$ or $(0,2)$ or $(1,2)$ in the future. The trail edge that was just introduced in the auxiliary graph offers connectivity in the graph to support only the request $(0,1)$ in the future. To include the additional reachability information, two additional edges $(V_0^{0,3}, V_1^{1,2})$ and $(V_0^{1,2}, V_1^{2,3})$ are introduced in the VT layers as seen in Figure 3.6(b). The three trail edges are identified with the same trail and the capacity of the edges is assigned to be the residual capacity of the trail. The properties (like capacity) of all the three edges are updated as more requests are set up or existing requests are torn down on the trail.

Assume that another connection $C_{1,2}^2$ arrives at the network. This request can be routed on the residual capacity of the existing trail T_1 and requires that a new transmitter on node 1 be activated. Since, the state of the transmission unit of node 1 on trail T_1 has changed, the trail edge $(V_0^{1,2}, V_1^{2,3})$ is removed and a new edge $(V_0^{1,3}, V_1^{2,3})$ is introduced. The residual capacity on all the trail edges are updated. Let us now consider two cases.

In the first case, $C_{1,2}^2$ leaves before $C_{0,2}^1$. The transmitter on node 1 is again idle on the trail. Hence, the edge $(V_0^{1,3}, V_1^{2,3})$ is removed, the edge $(V_0^{1,2}, V_1^{2,3})$ is added, making it consistent with equation (5) and the capacity on all the edges are updated. If $C_{0,2}^1$ leaves before another connection arrives, all the trail edges are removed and the original wavelength links are replaced.

In the second case, $C_{0,2}^1$ leaves before $C_{1,2}^2$, the wavelength link $(0,1)$ in the trail T_1 is free. Control plane signaling could be used to dimension the trail and T_1 is modified to $\{1,2\}$. The trail edges $(V_0^{0,3}, V_1^{2,3})$ and $(V_0^{0,3}, V_1^{1,2})$ are removed and the wavelength edge $(V_0^{0,1}, V_1^{1,1})$ is added as observed in Figure 3.7(a). Now, a new connection $C_{0,2}^3$ arrives. The connection can be carried from node 0 to node 1 using the just freed wavelength link $(0,1)$ creating a new trail $T_2 = \{0,1\}$, and then be electronically groomed by node 1 to be multiplexed along with trail T_1 to reach node 2. This activates the transmitter on node 0 and the receiver on node 1. Hence new trail edges $(V_0^{0,3}, V_1^{1,3})$ is added and the wavelength edge $(V_0^{0,1}, V_1^{1,1})$ is removed as seen in Figure 3.7(b).

3.5.5   Traffic Aggregation Penalties

Traffic aggregation techniques improve network utilization by providing more choices to accommodate incoming requests. Apart from the additional software and hardware that are required to perform aggregation, the o-grooming techniques generally incur a price which we call the aggregation/o-grooming penalty. There are two kinds of penalties - bandwidth penalty and transceiver penalty. We explain both the penalties through an example.

Consider a linear circuit A-B-C. Suppose that the circuit carries two connections - one from A to C and one from A to B. Also, suppose that the capacity of the wavelength is five units and that the bandwidth of each carried connection is 2 units.

The bandwidth penalty arises due to the reason that the connection from A to B locks up a bandwidth of 2 units on link (B,C) which is beyond the span of the connection. If o-grooming is used, this is a penalty that cannot be avoided and is called the bandwidth penalty.

Also, note that a total of three communication units are used - a transmitter on node A, a receiver each on nodes B and C. The total capacity of the communication units is 12 units. However, a wavelength is capable of filling only a maximum of 4 units of transmitter capacity and 4 units of receiver capacity. So, by tuning a second receiver into the wavelength, 4 units of receiver capacity is wasted. If, in general, there are x transmitters and y receivers in excess on an LT, the transceiver capacity lost (scaled by the wavelength capacity) is $(x + y)/2$.

By virtue of aggregation, even if packing of requests into circuits are done in the ideal way, transceiver utilization of 100 % is not achievable if more than one transmitter/receiver is tuned into a circuit and a wavelength utilization of 100% is not achievable if a connection that does not span the length of the trail is multiplexed into the trail.

3.6   Conclusions

In this chapter, we introduced a new metro network architecture called the PLATOON and reviewed the switch architecture of the constituent nodes. We decomposed the static single-hop network design problem into two subproblems - trail routing subproblem and wavelength assignment subproblem. We proved that the single-hop trail routing problem is NP-Complete and proposed three heuristics to solve it approximately and efficiently. We introduced the trail length constraint and discussed the two-hop static design problem in sparse grooming networks. We designed an auxiliary graph based approach to model multi-hop connection provisioning for PLATOONs with heterogenous network capabilities. The performance evaluation based on all the models described above is discussed in the Chapter 4.

CHAPTER 4. Network Dimensioning - Simulation Results

We evaluate the performance of the heuristics that we proposed in Chapter 3. We analyze the results for single hop PLATOONs following which we study two hop PLATOONs with sparse grooming capabilities. Finally, we discuss the results for dynamic provisioning in multi-hop PLATOONs with and without constrained client layer electronic speeds.

Throughout this chapter, we make the following assumptions. Every link in the network is assumed to have two fibers and the same wavelength can be used in the forward and reverse direction. Wavelength conversion is not present in the network. The traffic is primarily sub-wavelength.

We use the Waxman graph model for generation of random topologies. In the Waxman model, n nodes are randomly distributed over a rectangular grid indexed by integer coordinates. The edges are introduced between nodes s and d with a probability given by [122],

$$P(s,d) = \beta e^{-\frac{D(s,d)}{L\alpha}}$$

where D(s,d) is the Euclidean distance from node s to node d, L is the maximum distance between two nodes, and α and β are parameters in the range (0,1]. Large values of β results in graphs with higher edge densities and small values of α increase the density of short edges relative to long ones.

4.1   Single-Hop Network Design

We present results obtained from ILP that are described in Section 3.3.1 and from heuristics LS, BF and LG that are described in Section 3.3.3. CPLEX 8.1.0 was used to solve the integer program for the topology shown in Figure 4.1(a) and the heuristics were used to obtain results

109

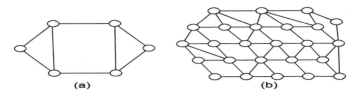

Figure 4.1   Test networks used for simulations (a) six-node network for ILP,
Diameter = 3 (b) 25 node network for heuristics, Diameter = 6

for the bigger network shown in Figure 4.1(b). The networks are assumed to use light-trail circuits to carry traffic. The capacity of a wavelength is arbitrarily assigned to be 48 units. A few of the variables used in this section are defined in Section 3.3.

4.1.1   Simulation Results

4.1.1.1   ILP Results

The requests between a node pair are of three granularities - 1, 3 and 12 units and number of such requests are uniformly distributed in the range (0,1), (0,1) and (2,3) respectively totaling an average offered traffic load of 925 units. Let T refer to the number of tunable transceivers per node, W, the number of wavelengths per link, and R, the total offered traffic. We can see from Figure 4.2 that when $T = 2$, throughput increases with W but saturates beyond $W = 8$. This is because there are not enough transceivers to set up the connections that are blocked. When T is increased to 3, all connections are accepted when $W = 5$.

4.1.1.2   Heuristic Results

Next, we present the TRAW heuristic results for the 25 node test network shown in Figure 4.1(b). The capacity of a wavelength is 48 units. The connections between any node pair are 1,3 and 12 units and the number of such connections are uniformly distributed in the range (0,12), (0,4) and (0,2) respectively. Figure 4.3(a) plots throughput as a function of number of tunable transceivers per node for the BF, LS and LG heuristics when 15 wavelengths are

T	W	Throughput(%)
2	6	81.41
2	7	82.38
2	8	83.14
2	12	83.14
3	4	92.21
3	5	100

Figure 4.2   ILP results for max throughput TRAW formulation on the six
node network shown in Figure 4.1(a) with aggregate traffic of
925 units

provisioned per link.

It is seen that a maximum of about 80 % throughput is obtained using any of the heuristics. Any increase in the number of transceivers does not help accept more connections. This is because the number of wavelengths in the link are not sufficient to establish the blocked connections. It is seen that BF heuristic curve raises much faster than the other two and saturates first. This suggests that the wavelength requirement for BF heuristic is high and the transceiver requirement is low. The performance of LS and LG heuristics are a little lower for lower values of T but saturates at a higher throughput value.

Figure 4.3(b) shows throughput as a function of number of wavelengths per link when 25 transceivers are provisioned per node. While BF heuristic achieves close to 90 % throughput, other heuristics saturate at about 78%. As observed earlier, the transceiver requirement for BF is low while for others, it is high. When the system is transceiver limited, BF outperforms other heuristics.

To achieve 100 % throughput for the current example when transceivers are not constrained, LS, BF and LG required 19, 22 and 17 wavelengths per link respectively. In general, the wavelength requirements of LG heuristic and transceiver requirements of BF heuristic were observed to be low for various scenarios. Since BF heuristic typically explores a large set of candidate trails, it is able to achieve better reduction in the number of transceivers. LG is a layered approach that considers routes other than shortest routes on the original network if

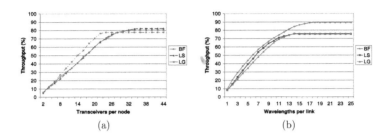

Figure 4.3   (a) Throughput as a function of number of transceivers per node
for 15 wavelengths per link. (b) Throughput as a function of
number of wavelengths per link for 25 tunable transceivers per
node

the corresponding wavelength is not available on the shortest route. This allows the first fit
wavelength allocation process to be more effective and helps conserve wavelengths.

In Figure 4.4(a), throughput is plotted as a function of number of wavelengths per link for
the fixed transceiver and tunable transceiver case. Since tunable transceivers are expensive,
the exact amount of savings generated by a tunable transceiver needs to be quantified to justify
their higher deployment costs. We quantify this savings with all the heuristics, but, we show
only the representative results using BF heuristics to prevent graph from getting cluttered.
An array of fixed transceivers contains a transmitter and receiver on every wavelength and
we use two such arrays per node. For the tunable case, we use 33 transceivers per node.
On an average, about 216 connections can be sourced by a node and hence the transceiver
requirement can be expected to be high. When 22 wavelengths per link are used, the number
of transceivers in the fixed case is 44 but is still unable to achieve 100 % throughput, while
using tunable transceivers, all the traffic are accepted.

Figure 4.4(b) plots the average utilization of a wavelength as a function of load for various
heuristics. We define the load as follows. For a node pair, every connection is established
corresponding to the above mentioned distribution with a probability p. We call p the load
of the system and it indicates the amount of traffic in the network. When load is 0.1, the
corresponding aggregate traffic is 1198 units and when load is 1, the aggregate network traffic

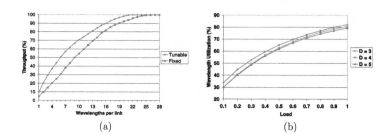

Figure 4.4 (a) Throughput as a function of number of wavelengths per link for tunable (35 per node) and fixed transceivers (2 arrays per wavelength per node) using BF heuristic. (b) Average wavelength utilization due to o-grooming as a function of load.

is 14,571 units in the example network. It is seen that BF heuristic has the lowest utilization and hence requires more wavelengths to accept all traffic. LG heuristic shows the best performance and about 90% of the wavelength is utilized at maximum load. Such high utilizations in the presence of fractional traffic leads to significant wavelength savings.

4.1.2 Conclusions

We formulated an ILP and proposed three heuristics for design of LT networks with limited wavelength and transceiver resources. We obtained optimal solutions for the design problem based on our ILP formulation. We observed that our heuristics lead to excellent wavelength utilization of upto 90 % even in the presence of heavily fractional traffic subject to non-bifurcation constraints. We quantified the savings achieved using tunable transceivers as compared with fixed transceivers.

4.2 Two Hop Sparse Grooming network design

A few of the variables used in this section are defined in Section 3.4.

N_h/W	3	2	1
0	389	365	292
1	413	402	306

(a)

	N_g	N_λ
LP + FG	6	3
LP + SG	1	3
LP + NG	0	4

(b)

	N_h	N_λ
LT + NH	0	3
LT + SH	1	3
LT + FH	6	3

(c)

Figure 4.5   Results from ILP for the six-node network (a) Maximize throughput formulation with $\delta = 2$ (b) Minimize cost formulation for lightpaths. LP = lightpaths, SG = sparse grooming, FG = full grooming, NG = no grooming (c) Minimize cost formulation for light-trails. LT + NH = light-trails with no trail length constraints and no hubs, LT + SH = light-trails with sparse hub nodes, $\delta = 2$, LT + FH = light-trails with all nodes hub capable, $\delta = 2$

4.2.1   Simulation Results

In this section, we present the results for ILP that was formulated in Section 3.4.2 and for heuristics PC, RC and EC that were presented in 3.4.3. The networks are assumed to use light-trail circuits to carry traffic. The capacity of a wavelength is arbitrarily assigned to be 48 units.

4.2.1.1   ILP Results

The ILP formulation in the max throughput form and the min cost form were solved using CPLEX 8.1.0 for the network shown in Figure 4.1(a). The requests between any node pair are of three granularities - 1, 3 and 12 units. With probability p, a connection of each granularity is established between a node pair and we set p = 0.5 for the illustrative example in Figure 4.5. The number of such 1, 3 and 12 streams are uniformly distributed between (0,1), (0,1) and (2,3) units respectively.

For the throughput maximization ILP problem in light-trails with $\delta = 2$, if $N_h = 0$ is added as an additional constraint, only 389 units are carried for W=3 or more as shown in Figure 4.5(a). If we set N_h=1, all the traffic (413 units) can be supported for W = 3 or more. At W=1, only 306 units (about 74%) of the traffic is carried even when hub capable nodes are present in the network showing that wavelength is the bottleneck.

For the cost minimization ILP problem, we assume $C_\lambda=1$ and $C_h = 1$, and provide all possible paths as input. We find that lightpaths with no grooming requires 4 wavelengths whereas lightpaths with full grooming require 3 wavelengths as shown in Figure 4.5(b). Lightpaths with sparse grooming requires only one grooming node and needs three wavelengths. This shows that sparse grooming can achieve performance close to full grooming. When light-trails with no trail length constraints are studied, no hubbing was required and still only 3 wavelengths were consumed. When the trail length limit constraint is imposed and trails of length 2 are provided as input, all the traffic is still carried using 3 wavelengths while one node is designated as a hub node as shown in Figure 4.5(c).

4.2.1.2 Heuristic Results

We apply our heuristics to study the effect of sparse hubbing on a 25-node light-trail network shown in Figure 4.1(b). The number of streams of sizes 1, 3 and 12 between a node pair are uniformly distributed as (0,12), (0,2) and (0,1) respectively. We set $W = 13$, $\delta = 3$, and observe average throughput of the network as a function of the number of H-nodes in Figure 4.6(a) by running the simulation with 500 different traffic matrices having the above distribution. When $N_h = 0$, about 74 % of the traffic is carried while the rest are physically blocked. With only a few H-nodes, the throughput climbs steeply and reaches close to 100 %. We observe that the PC heuristic yields the best throughput closely followed by the EC heuristic. The RC heuristic performs the worst but they all converge to the same value as the number of H-nodes increase.

We study the average wavelength requirement by running the simulation for 2000 traffic matrices of above distribution in Figure 4.6(b). In this case, if we do not observe 100 % throughput, it is because of physically blocked node pairs and not because of wavelength exhaustion. We observe that EC heuristic yields the minimum number of wavelengths closely followed by the PC heuristic. The random heuristic, on an average, is unable to achieve 100 % throughput until about $N_h = 11$ (not shown in figure), while the other two yield 100 %

115

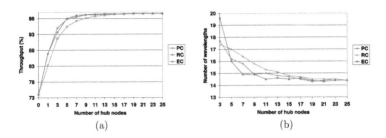

(a)   (b)

Figure 4.6   (a) Network throughput as a function of the number of hub
nodes for the 25 node network with W = 13, $\delta = 3$ (b) Wavelength requirements for 100 % throughput (throughput is less
for RC heuristic) at $\delta = 3$, as a function of the number of hub
nodes for the 25 node network

throughput with only just one hub node. Since the hubs are randomly chosen in RC heuristic,
the chosen hubs may not be in the proximity of every physically blocked node pair. In both the
PC and EC heuristic, the first node that is chosen corresponds to the center of the graph. Since
the diameter is 6, and the center of the graph has the longest path only of 3 from it, the first
chosen vertex is in the proximity of every physically blocked pair. If wavelength availability
is not a bottleneck, it can serve as the hub for all physically blocked pairs. Hence, 100 %
throughput can be observed even with one hub node.

As wavelength requirements decrease with increase in number of hub nodes, it may be
interesting to identify the exact number of hub nodes required for a given traffic scenario. We
can identify the network cost similar to the approach in [147].

Define the ratio ρ,

$$\rho = \frac{C_h}{C_\lambda}$$

The cost of the network C_n is

$$C_n = N_h \times C_h + N_\lambda \times C_\lambda = (N_h \times \rho + N_\lambda) \times C_\lambda$$

Normalizing the cost by C_λ,

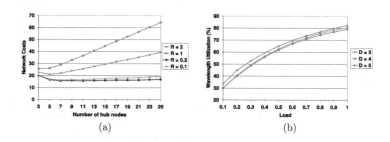

Figure 4.7   (a) Network costs plot to determine the optimal number of hub nodes (b) Wavelength utilization as a function of load for varying trail length limits (D)

$$C_n = N_h \times \rho + N_\lambda \qquad (4.1)$$

Figure 4.7(a) shows the cost of the network for various values of ρ. If the cost of the hub node is much larger than maintaining a wavelength, it may be better to operate with minimal number of hub nodes. For this specific example, if $\rho = 1$ or 2, the optimal cost is achieved at N_h=5 while for $\rho = 0.1$ or 0.2, the optimal cost is at N_h=7. The utilization of a wavelength for various values of trail sizes are shown in Figure 4.7(b) . The load on x-axis corresponds to the parameter p described above. As p increases from 0 to 1, it can be seen that the wavelength utilization steadily increases and reaches about 82 % indicative of good packing by our heuristic. Heavier load at shorter δ is due to traffic rearrangement.

4.2.2   Conclusions

We studied the sparse hubbing problem in light-trail networks. We adopted a unified approach for ILP formulation that is applicable for both groomed-lightpath and hubbed-light-trail networks, and presented results for a test network. We designed simple heuristics for H-node placement, traffic rearrangement and light-trail routing in the context of multiple granularity connections subject to non-bifurcation constraints. Our simulation results suggest that with only a small number of hub nodes, high network throughput and good wavelength utilization can be achieved. Our research also gives guidelines for deciding the network operation point based on network element costs.

4.3   Multi-hop network dimensioning

In this section, we define some of the performance metrics of interest and present the simulation results obtained for the auxiliary graph based heuristic that was described in Section 3.5.4. A few of the variables used in this section are defined in Section 3.5.

Throughout this section, call arrivals are poisson distributed and call holding times are exponentially distributed (whose average value is normalized to unity). Traffic is uniformly distributed among all node pairs. The capacity of the wavelength is arbitrarily assigned the value 10. The simulations are done using discrete event simulation techniques for a series of 10^6 call arrivals (unless explicitly stated otherwise). The observed results are averaged and reported in this section.

4.3.1   Performance Metrics

The metrics of interest can be broadly categorized into three types : connection related, network related and circuit related.

Connection Related:

- Average number of connections

- Average physical hop length per connection

- Average virtual hop length per connection

- Connection blocking probability per granularity

- Connection blocking probability

- Capacity blocking probability

- Service Provisioning Time

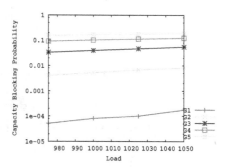

Figure 4.8   Effect of granularity on blocking performance with W = 17 and X = 21

Network Related:

• Wavelength Related

    – Average number of wavelengths links used

    – Average total wavelength bandwidth used

    – Wavelength packing fraction

    – Link wise distribution of capacity used

• Transceiver Related

    – Average total number of transceivers used

    – Average capacity of transceivers used

    – Transmitter to receiver usage ratio

    – Node wise Distribution of transceiver usage

Circuit Related:

• Average number of circuits

• Average number of connections per circuit

• Average circuit length

119

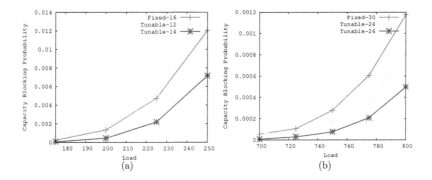

Figure 4.9   Effect of tunability on blocking performance(a) Throughput as
a function of W for X = 21 and L = 450E (b) Throughput as a
function of X for W=27 and L=450E

Let η be the metric which holds the value η_i during interval t_i which is the time period
between the ith event (connection arrival or departure) and $(i+1)^{th}$ event. Let T be the total
time of simulation. The average value of η is given by

$$\eta_{avg} = \frac{\sum_i \eta_i * t_i}{T}$$

4.3.2   Single-Hop Results

The results reported in this section assume that electronic grooming capabilities are not
available in the client layer. We study the impact of granularity and tunability on blocking
performance of LT networks. We also compare SLT, DLT, LT and LP in terms of various
performance metrics.

4.3.2.1   Effect of Granularity

In this section, we study the impact of granularity on blocking performance in LT networks.
Let Gi be the maximum granularity request for scenario i, with i varying from 1 to 5. To enable
fair comparison, the network is loaded to equal extent for each scenario and the capacity

Network	α	β	Nodes	Links	Diameter
N1	0.4	.25	20	63	5
N2	0.4	.15	30	90	7
N3	0.4	.1	40	121	8

Figure 4.10   Parameters for the random graphs that were used for the simulations

blocking performance is compared for each of these instances. Let N_i be the number of call arrivals at the network for scenario i. The capacity arrival rates of individual granularities are always equal for any scenario. So, for a scenario with granularity Gi, the probability that a given call is of size k is given by

$$p_k = \frac{1}{k} \frac{1}{\sum_{j=1}^{i} \frac{1}{j}}$$

Hence, the average size of an incoming call for scenario i is s_i and is given by:

$$s_i = \frac{i}{\sum_{j=1}^{i} \frac{1}{j}}$$

In our simulations, for scenario G1, $s_1 = 1$ and the network is loaded with $N_1 = 10^6$ arrivals. For scenario Gi, the network is loaded with N_i arrivals where N_i is given by

$$N_i = s_i * N_1$$

This ensures that in all the problem instances, the network is loaded to equal extent. We performed simulations on a 20 node random network with 39 bidirectional links, diameter of 5 and an average path length of 2.5. The number of wavelengths were varied from 14 to 17 and the number of transceivers were varied from 18 to 21. The blocking probability shown in Figure 4.8 is for 17 wavelengths and 21 transceivers. As the number of granularities increase, the blocking performance degrades steadily. It can be noted that, there are about four orders of magnitude difference between instances G1 and G5, showing that number of granularities of traffic has a major effect on network performance.

Network	W	X	R
N1	20	15-19	450-550
N2	27	16-20	450-550
N3	21	15-19	450-550

(a)

Network	W	X	R
N1	15-19	20	450-550
N2	18-26	21	450-550
N3	13-21	20	450-550

(b)

Figure 4.11   Provisioned resources for (a) transceiver limited scenario (b) wavelength limited system

4.3.2.2   Effect of Tunability

We study the impact of tunability on network performance. Tunable transceivers are expensive but they are more versatile when the exact call arrival and departure events are not known in advance. We study the blocking performance as a function of load for a 20 node random network with 39 bi-directional links and 8 wavelengths on each link in the presence of fixed and tunable transceivers. The load is varied from 175 to 250 Erlangs. It is seen in Figure 4.9(a) that a network with 14 tunable transceivers per node can perform better than a network with 16 fixed transceivers per node. The number of wavelengths is increased to 16, and load is varied from 700 to 800 Erlangs as shown in Figure 4.9(b). In this scenario, a network with 24 tunable transceivers is able to perform much better than a network with 30 fixed transceivers. In general, as load increases, tunable transceivers show more improvement in performance as compared with fixed transceivers.

4.3.2.3   Comparative Study

In this section, we compare performance of LP, SLT, DLT and LT architectures in random mesh networks that were generated based on the Waxman model mentioned in the beginning of the chapter. We studied three random networks - N1, N2 and N3, with 20, 30 and 40 nodes respectively. Since the results obtained in each case were similar in nature, we report it only for the N2 network. The network has two resources - wavelengths and transceivers. For each topology, we study two kinds of scenarios - wavelength limited scenario and transceiver limited scenario. For a wavelength (transceiver) limited scenario, we provide plenty of transceivers

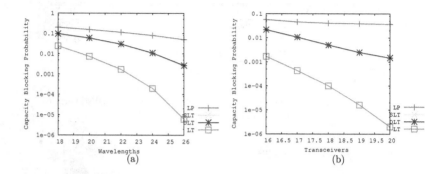

Figure 4.12   (a) Throughput as a function of W for X = 21 and L = 450E
(b) Throughput as a function of X for W=27 and L=450E

(wavelengths) to ensure that the number of wavelengths (transceivers) alone is critical for network performance. The Waxman parameters used for generating the three random networks are provided in Figure 4.10. The number of wavelengths per link (W) and number of transceiver pairs per node (X) provisioned in each network are reported in Figure 4.11.

4.3.2.4   Capacity Blocking

The capacity blocking probability metric (B) is plotted as a function of W in Figure 4.12(a) with X set to 21. The load (R) offered to the network is the average arrival rate since the average holding time is unity, and is set to 450 Erlangs. It is observed that as wavelength increases, there is a significant improvement in blocking performance for all networks. The improvement is more for LT networks than for the others. LT shows the best performance while LP shows the worst performance. SLT and DLT show similar performance and is positioned between LT and LP. This is due to the reason that LT has more aggregation choices than SLT and DLT which in turn have more choices than LP. SLT and DLT have identical aggregation choices and hence they show similar performance. In Figure 4.12(a), when W=18, by adding only 8 more wavelengths per link, a performance improvement by about 4 orders of magnitude is observed for LT. For the rest of the comparative study in this section, we consider only transceiver

123

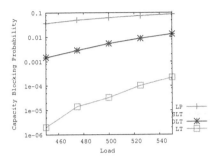

Figure 4.13 Blocking performance as a function of load for X = 20 and W
= 27

constrained systems.

The blocking performance as a function of number of transceivers is reported in Figure
4.12(b) with W set to 27 and R set to 450 Erlangs. Here again, LT performs the best, LP
performs the worst, while SLT and DLT perform somewhere in between. There exists about
four orders of magnitude difference between LT and LP in both transceiver and wavelength
limited systems. In Figure 4.12(b), when X=16, by adding only four more transceivers, the
blocking performance improves by about four orders of magnitude. This suggests that blocking
is more sensitive to increase in transceivers than to increase in wavelengths.

The blocking performance as a function of load for X = 20 and W = 27 is seen in Figure
4.13. As load increases, there is a gradual increase in blocking. The blocking performance is
observed to be more sensitive to change in loads at low loads than at high loads. We report
below a few more metrics that were studied as a function of load, with W = 27 and X = 20.

4.3.2.5 Circuit and Connection Lengths

The average circuit length is plotted as a function of load in Figure 4.14(a). The average
path length of this network is 3.17. In general, the average circuit length increases with load and

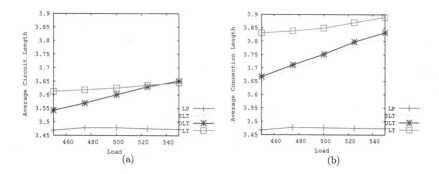

Figure 4.14   (a) Average Circuit Length as a function of load when X = 20
and W = 27 (b) Average Connection Length as a function of
load when X = 20 and W = 27

saturates at some load. As load increases initially, circuits with longer lengths are established
since wavelengths become unavailable along shorter routes. However, as load increases further,
fewer connections get accepted due to increase in blocking and hence the increase in circuit
lengths is more subtle. When all wavelength links are exhausted, the network stops accepting
connections and hence the circuit lengths saturate. For low offered loads, the carried load is
maximum for LT due to its increased aggregation choices and hence LT has the highest average
circuit length. However, beyond a load of about 520 Erlangs, SLT and DLT have longer circuit
lengths than LT. This is because SLT and DLT carry less traffic at low offered loads but at high
loads, due to reduced aggregation choices, longer circuits are required to carry more traffic.
LP network is saturated for the studied offered loads.

The average connection length is examined as a function of load in Figure 4.14(b). The
trends in the average connection length are similar to those observed for average circuit lengths.
The average connection lengths are greater than average circuit lengths. The average connection length for LT is approximately about 20 % in excess of the average path length while for
SLT and DLT, it is about 15 % in excess of the same. This can be explained by the fact that
average connection length can be defined as follows:

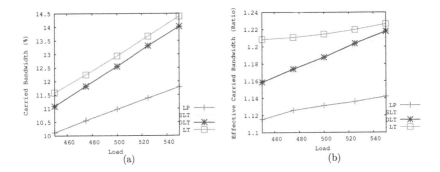

Figure 4.15   When X = 20 and W = 27, (a) carried bandwidth as a function
of load (b) effective carried bandwidth as a function of load

$$CL_{avg} = \frac{\sum_i TL_i * C_i}{\sum_i C_i}$$

where CL_{avg} is the average connection length, TL_i is the length of trail i and C_i is the number of connections in trail i. This expression states that the average connection length can be computed by picking each trail length and weighing it by the number of connections in the trail. Since longer trails are likely to have more connections, they are likely to be weighted more and hence average connections lengths are greater than average circuit lengths.

4.3.2.6   Carried Bandwidth

Carried bandwidth is defined as the total bandwidth consumed to carry network traffic. For a specified carried load, the lesser the required carried bandwidth, the better the performance. Figure 4.15(a) plots carried bandwidth normalized to the total bandwidth in the network as a function of load. LT carries maximum load due to its increased aggregation choices. Since the network load is high, individual circuits take longer routes and hence the bandwidth consumed to carry the given traffic would also be high. LP, on the other hand, carries the least load and hence consumes the least network bandwidth. It is important to note that with only a small

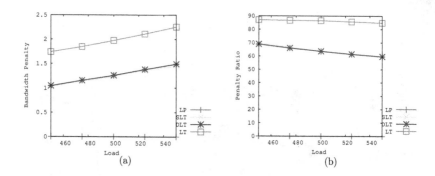

Figure 4.16 (a) Bandwidth aggregation penalty as a function of load when
X = 20 and W = 27 (b) Penalty ratio as a function of load for
X = 20 and W = 27

increase in bandwidth consumption, LT is able to achieve a very significant improvement in performance. For instance, at 450 Erlangs, LT and LP consume about 11.5 % and 10 % of the total bandwidth in the network respectively while SLT and DLT consume about 11 % of the total bandwidth. However, LT is able to achieve a performance improvement of four orders of magnitude as compared with LP with this small increase in bandwidth consumption as shown in Figure 4.12(b). As offered load increases, the carried load also increases, thereby increasing required carried bandwidth.

Effective carried bandwidth is the amount of required carried bandwidth normalized by the minimum required bandwidth to carry a specified offered load. To calculate the minimum required bandwidth for a specified load, it is assumed that every call is routed along the shortest route and there are no wavelength or transceiver limitations. Effective carried bandwidth is plotted as a function of load in Figure 4.15(b).

It is seen that the bandwidth consumed is only about 20 % more than the minimum required value to carry the load. Here again, LP consumes the least bandwidth in excess of the minimum bandwidth because it has the least carried load. For all architectures, as load increases, deviation from the minimum value increases.

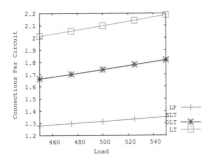

Figure 4.17   Number of connections per circuit as a function of load when
X = 20 and W = 27

The bandwidth required to carry a connection can be logically construed to be of three parts: B_{min}, B_p, and B_{cr}. B_{min} is the minimum required bandwidth to carry the connection assuming that there are no resource constraints. B_p is the bandwidth consumed due to multiplexing penalty which arises because of the nature of aggregation technique used and is discussed in Section 3.5.5. It is integral to aggregation and has significant effect on network performance. The third component refers the to fact that the connection may not always follow the shortest route due to the limited availability of resources along that route or due to the routing policy.

Consider an example linear circuit $A1 - A2 - A3 - A4 - A5$, which among other connections, carries a connection of unit size from $A1$ to $A4$. Also, assume that there exists a link from $A1$ to $A3$ on the physical topology. Here $B_{min} = 1$, since any connection from $A1$ to $A3$ will require a minimum of one hop. Since, $A1$ and $A3$ are separated by two links in the selected circuit, an excess of one hop was required due to constrained resources along the shortest route (or due to some routing policy) and so $B_{cr} = 1$. Since the connection locks up bandwidth on the links - $(A3,A4)$ and $(A4,A5)$, $B_p = 2$.

128

4.3.2.7 Bandwidth Penalty

The bandwidth penalty, normalized to the total carried capacity, is plotted as a function of load in Figure 4.16(a). LP supports only source-destination aggregation and hence does not have the multiplexing penalty. The multiplexing penalty is a minimum of 15 % in LT networks while it is a minimum of 9 % in SLT/DLT networks. Since, penalty is a significant per cent of the total carried bandwidth, it plays an important role in the performance. As load increases, the total carried capacity also increases. The penalty also increases due to increased multiplexing such that the ratio of bandwidth penalty to carried capacity increases.

Penalty ratio is defined as the ratio of B_p to $B_p + B_{cr}$, expressed as a %. It is seen that the ratio is as high as 85 % for LT in Figure 4.16(b). This suggests that, in LT, of the network bandwidth required in excess of the minimum required bandwidth, bandwidth penalty has much greater role to play than constrained resources or routing policy. Since the aggregation choices are limited for SLT and DLT, lesser aggregation happens and hence the bandwidth penalty is only up to a maximum of 70 %. As load increases, the bandwidth penalty increases. However, the bandwidth consumed due to constrained resources (or routing policy) increases even more rapidly due to higher carried load. Hence, on the whole, penalty ratio decreases with increase in offered load.

4.3.2.8 Multiplexing Gain

Multiplexing gain refers to the average number of connections sharing a circuit. It is seen in Figure 4.17 that LT has the maximum number of connections multiplexed per circuit, followed by SLT and DLT , with LP having the least multiplexing capability. It is also observed that as load increases, sharing ability increases. At higher loads, when a circuit is set up, there are good chances of another call arriving at the network to be multiplexed on to the same circuit with the calls arriving at a more rapid rate.

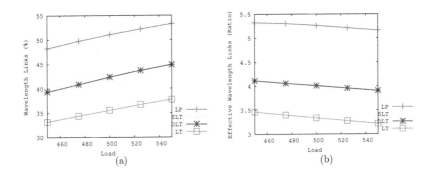

Figure 4.18   (a) Required wavelength links expressed as a % of the total
wavelength links in the network as a function of load when X
= 20 and W = 27 (b) Effective wavelength links as a function
of load for X = 20 and W = 27

4.3.2.9   Wavelength Links Usage

Effective carried bandwidth alone is not a good metric for evaluating the efficiency of
grooming because it does not specify how the bandwidth usage is distributed. A metric that
captures the information of how fractional the packing is the number of wavelength links used.
A wavelength is said to be *touched* if it is being used for carrying some traffic. The number of
wavelength links used as a % of the total number of wavelength links in the network is shown
in Figure 4.18(a). For a specific carried load, the lower the number of wavelength links used,
the better the performance is.

It is seen that as load increases, the number of used wavelength links increases for all
architectures. LT uses the minimum number of wavelength links despite the fact that it
carries the maximum load as compared with other architectures. LP uses the most number of
wavelength links and the performance of SLT/DLT is somewhere in between LP and LT. It is
inferred from Figure 4.15(a) and 4.18(a) that LP touches about 50 % of the wavelength links
in the network to carry bandwidth of just over 10 % of total capacity in the network showing
that the packing is very fractional in nature.

130

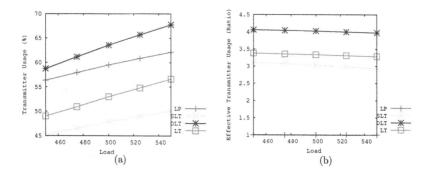

Figure 4.19 (a) Transceiver usage expressed as a % of the total number of
transceivers in the network as a function of load when X = 20
and W = 27 (b) Effective transmitter usage as a function of
load for X = 20 and W = 27

A loose lower bound on the required number of wavelength links is the number of wavelength
links required to carry the minimum carried bandwidth which is when all the traffic traverses
the shortest routes.

$$WL_{min} = \lceil \frac{\sum_{i,j} T_{i,j} * d_{i,j}}{C} \rceil$$

where, WL_{min} refers to the minimum wavelength links required to carry the traffic, $T_{i,j}$
is traffic from i to j, C is the capacity of a wavelength and $d_{i,j}$ is the shortest distance
between nodes i and j. The required number of wavelength links normalized to the minimum
required wavelength links is effective wavelength links and is plotted as a function of load in
Figure 4.18(b). It is seen that LT consumes about 500 % in excess of the minimum required
wavelength links. Please note that the deviation is large only because the bound is loose and
is not practically realizable. The deviation from the minimum value decreases with increase
in load. This is because the aggregation capability increases with load and consequently
new traffic is accommodated more on existing circuits and not by always opening up new
wavelengths.

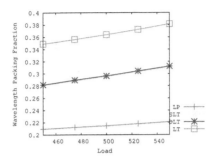

Figure 4.20 Wavelength packing fraction as a function of load when X = 20 and W = 27

4.3.2.10 Wavelength Packing Fraction

A metric that combines both carried bandwidth and the number of wavelength links used is the wavelength packing fraction. This metric specifies the fraction of a touched wavelength that is actually used to carry traffic. For a given load, in single-hop networks, the more the packing fraction, the better the performance. The packing fraction of a used wavelength as a function of load is reported in Figure 4.20. For LT, about 35 % of a wavelength is packed while for LP, a little over 20 % is packed for the given range of load. This metric suggests that, for an offered load, not only is LT able to accommodate more requests but also it is able to pack it efficiently without using a lot of wavelengths. The packing fraction increases as load increases since multiplexing capability increases with increasing load.

4.3.2.11 Transmitter Usage

Transceivers are expensive equipment in an optical network and it is important to compare the transceiver requirements of the various architectures. Transmitter usage is defined to be the number of transmitters touched normalized by the number of transmitters in the network. For a given carried load, the lesser the transmitter usage, the better the performance is. For a specified offered load, SLT has the minimum transmitter usage since it uses a source based

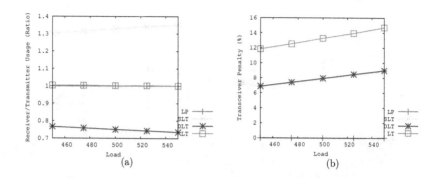

Figure 4.21   (a) Transceiver usage expressed as a % of the total number of transceivers in the network as a function of load when X = 20 and W = 27 (b) Effective transmitter usage as a function of load for X = 20 and W = 27

aggregation technique as observed in Figure 4.19(a). DLT, on the other hand, uses only a destination based aggregation technique and uses transmitters rapidly thereby leading to highest transmitter usage. LP, with its source-destination aggregation performs a little better than DLT. Though LT has more aggregation choices than SLT, LT tries to conserve both transmitters and receivers and hence performs slightly worse than SLT.

A loose lower bound on the required number of transmitters can be computed as follows:

$$T_{min} = \sum_i \lceil \sum_j \frac{T_{i,j}}{C} \rceil$$

where X_{min} is the minimum number of transmitters required in the network. The total number of used transmitters normalized to the minimum required transmitters is the effective number of transmitters and is plotted in Figure 4.19(b). It is seen that DLT deviates the most from the minimum and is about 400 % in excess of the minimum required. SLT, which performs the best is about 300 % in excess of the minimum. The effective number of transmitters reduces with load because the increased grooming capability at higher loads leads to more efficient consumption of transmitters.

While DLT performs poorly in transmitter consumption, it achieves the best receiver con-

Network	α	β	Nodes	Links	Diameter
N1	0.4	.25	20	63	5
N2	0.4	.15	30	90	7
N3	0.4	.1	40	121	8

Figure 4.22   Parameters for the random graphs that were used for the simulations

sumption since it performs destination level aggregation. The ratio of transmitter to receiver usage is plotted for the various architectures in Figure 4.21(a). It is observed that SLTs use 30 % more receivers than transmitters and DLTs use 25 % more transmitters than receivers. For SLT, the ratio increases with load. This is because as load increases, due to increased aggregation, transmitter usage does not increase as much as the receiver usage.

4.3.2.12   Transceiver Penalty

The transceiver penalty (TP) scaled by the total transceiver capacity available in the network is evaluated as follows :

$$TP = \frac{100 * \sum_{t}^{T} \left(N_{tx}^{t} + N_{rx}^{t} \right) - 2 * T}{2 * X}$$

where N_{tx}^{t} is the number of transmitters on the tth trail, N_{rx}^{t} is the number of receivers on the tth trail, T is the number of trails, and $2 * X$ is the sum of the number of transmitters and receivers in the network. TP is plotted as a function of load in Figure 4.21(b). Transceiver penalty increases with load due to increased aggregation capabilities at higher load. Due to transceiver penalty, about 12 % of the total transceiver capacity is lost for LT. About 7 % of the total transceiver capacity is lost for SLT and DLT. This corresponds to transmitter capacity loss in DLT and receiver capacity loss in SLT. LP does not suffer from transceiver penalties since each circuit has exactly one transmitter and one receiver.

4.3.2.13   Single-Hop Summary

The path level aggregation techniques have significant impact on blocking performance. Traffic aggregation, by the nature of its implementation, leads to bandwidth and transceiver penalties. However, LT performs several orders of magnitude better than other architectures. SLT and DLT have similar blocking performance but SLT consumes more receivers while DLT consumes more transmitters. If transmitters are more expensive than receivers, performance of SLT can be improved by asymmetrically provisioning more receivers on the node than transmitters. LP shows the worst blocking performance,but has the important advantage of simplicity.

4.3.3   Multi-Hop Results

The auxiliary graph approach is used to compare network performance under both network interconnection models - overlay model and integrated model. For the overlay model, the client layer does not have any visibility into the optical layer. When a call arrives, it is first routed on the virtual topology. If the call can be accommodated, a request is made to the optical layer where if there are sufficient wavelength links available, the call is accepted. Otherwise, the call is rejected. For the integrated model, the client layer has complete visibility into the optical layer. When a call arrives, decisions are taken taking into account state in the virtual topology and in the optical layer. Two e-grooming policies were used for the integrated model - one that minimizes the number of virtual hops taken by a connection and the other that minimizes the number of physical hops taken by a connection.

For both the models, eight architectures - LP, SLT, DLT, LT, LP-TG, SLT-TG, DLT-TG and LT-TG were compared. We report results only for the overlay model since the derived conclusions were similar in nature for both the overlay model and peer model. Simulations were run for three random network topologies - N1, N2 and N3 with 20, 30 and 40 nodes respectively. The Waxman parameters for these random topologies are provided in 4.22. The number of transceivers and wavelengths provisioned for the network are specified in Figure 4.23. For each topology, the wavelength limited and transceiver limited scenarios were studied.

Network	W	X	R
N1	19	12-16	450-550
N2	20	10-14	450-550
N3	19	14-18	450-550

(a)

Network	W	X	R
N1	12-16	20	450-550
N2	14-19	18	450-550
N3	12-16	25	450-550

(b)

Figure 4.23   Provisioned resources for (a) transceiver limited scenario (b) wavelength limited system

We describe the results obtained for the 40 node N3 network here since the other topologies yielded similar conclusions.

4.3.3.1   Capacity Blocking

The capacity blocking performance as a function of X is shown in Figure 4.24(a) for a transmitter limited system with W = 19 and R = 450 Erlangs. The blocking performance as a function of W for a wavelength limited system with X = 25 and R = 450 Erlangs is reported in Figure 4.24(b). We make the following observations from the figures.

- Architectures with e-grooming capabilities outperform other architectures by multiple orders of magnitude.

- For transceiver limited systems, LP-TG outperforms LT-TG, SLT-TG and DLT-TG by many orders of magnitude.

- For wavelength limited systems, LT-TG, SLT-TG and DLT-TG outperform LT-TG by many orders of magnitude.

- For both wavelength and transceiver limited systems, SLT-TG, DLT-TG and LT-TG show similar performance.

We can explain the observations made above as follows. For a circuit passing through N nodes, the number of edges introduced in the virtual topology is $O(1)$ for LP-TG, $O(N)$ for SLT-TG and DLT-TG and $O(N^2)$ for LT-TG. Since the number of edges introduced is

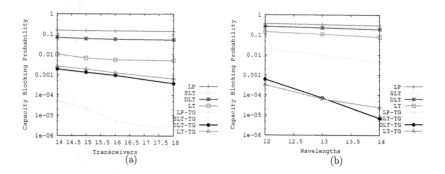

Figure 4.24 (a) Capacity blocking probability as a function of X with W = 19 and Load = 450E (b) Capacity blocking probability as a function of W with X = 25 and Load = 450E

maximum for LT-TG, the density of the virtual topology layer is the highest for LT-TG and the lowest for LP-TG. Due to the low density of LP-TG virtual topology layer, its diameter is large and the connectivity is low. In the overlay model, the route is first found on the virtual topology layer. Due to low reachability, the route that is found takes lot of hops to reach the destination and hence may require lot of wavelength links. In wavelength limited systems, consuming a lot of wavelength links is detrimental to performance and hence LP-TG performs poorly.   To corroborate this reasoning, we observe average connection length as a function of load in Figure 4.25(a). It is seen that for a given load, the average connection lengths assumed by LP-TG is about 40 % higher despite carrying lesser load than other e-grooming architectures. LT-TG, SLT-TG and DLT-TG have similar average connection lengths but lower than LP-TG. The number of used wavelength links scaled to the total number of wavelength links in the network is observed as a function of load in Figure 4.25(b). It is noted that LP-TG consumes about 50 % more wavelength links than other architectures with e-grooming capabilities despite carrying lesser load. This confirms that LP-TG experiences a more fractional packing and longer routes due to the sparse virtual topology. Non e-groomed architectures consume more number of wavelength links than e-groomed architectures due to

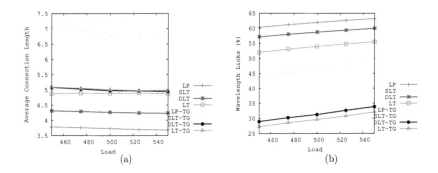

Figure 4.25   For wavelength constrained scenario: (a) Average connection
length as a function of load with X = 25 and W = 15 (b)
Required wavelength links as a % of the total wavelength links
in the network as a function of load with X = 25 and W = 15

their heavily fractional packing tendencies. The average connection lengths for e-groomed
architectures are larger than their purely o-groomed counterparts because e-grooming leads to
lower blocking and acceptance of more connections with long routes.

To understand why SLT-TG, DLT-TG and LT-TG systems perform poorly in transceiver
constrained systems, we look at the call routing algorithm for the e-groomed architecture. In
overlay systems, the virtual topology is first searched to route the call. Due to significant
density of the virtual topology in SLT-TG, DLT-TG and LT-TG, chances of find an edge
between the source and the destination of the call is high. However, the communication units
on the source and destination may not be already provisioned to route the call. We described in
Section 3.5.5 that whenever more than one transmitter or receiver is multiplexed on a circuit,
transceiver penalty is incurred. In transceiver constrained systems, penalty on an already
precious resource leads to LT-TG, SLT-TG and DLT-TG performing poorly.

To explain why SLT-TG, DLT-TG and LT-TG show identical performance, it is to be noted
that improvement in performance due to e-grooming dominates the performance improvement
due to o-grooming making o-grooming gains as a second order effect. LT-TG has more choices

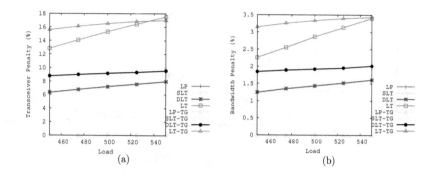

Figure 4.26 For transceiver constrained scenario: (a) Transceiver penalty
as a function of load with W = 19 and X = 15 (b) Multiplexing
bandwidth penalty as a function of load with W = 19 and x
= 15

in o-grooming than SLT-TG and DLT-TG. However, the bandwidth and transceiver penalties are also increased due to increased o-grooming choices in LT-TG. The improvement in
performance due to increased o-grooming is offset by the bandwidth and transceiver penalties
thereby leading to similar performance for all the three architectures.

To corroborate this reasoning, we study transceiver penalty as a function of load in Figure
4.26(a). It is noted that the transceiver penalty of LT-TG is about 100 % more than what
is observed for SLT-TG and DLT-TG. At low loads, LT shows lower transceiver penalty than
LT-TG, but it overshoots LT-TG at high loads. The rapidly increasing penalty trend in LT
is due to its increased multiplexing tendency with increasing loads. There are more number
of ways in which a call can be accommodated in a network with e-grooming capabilities. Due
to this, the possibility of multiplexing an extra communication unit on an existing circuit is
lower, thereby leading to a more gradual increase in transceiver penalty as compared with the
non e-groomed LT. The bandwidth penalty as a function of load is seen in Figure 4.26(b). The
bandwidth penalty incurred by LT-TG is about 70 % more than SLT-TG and DLT-TG.

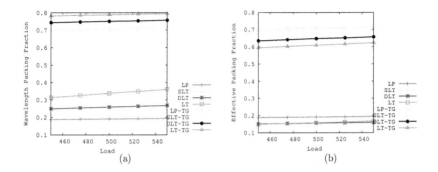

Figure 4.27   For transceiver constrained scenario: (a) Wavelength packing fraction as a function of load with W = 19 and X = 15 (b) Effective wavelength packing fraction as a function of load with W = 19 and x = 15

4.3.3.2   Wavelength Packing Fraction

The wavelength packing fraction of the transceiver constrained system as a function of load is plotted in Figure 4.27(a). It is seen that LT-TG has a higher wavelength packing fraction than SLT-TG, DLT-TG and LP-TG. It appears that LT-TG should perform better than LP-TG. However, it is important to note that this metric does not account for the fact that some parts of the carried bandwidth in a wavelength is actually not useful and is simply bandwidth penalty. We subtract the bandwidth penalty from used bandwidth to study the effective wavelength packing fraction which is shown in Figure 4.27(b). It is seen that when useful bandwidth alone is considered, LP-TG performs better than LT-TG, DLT-TG and SLT-TG and LT-TG performs the worst. This clearly shows that bandwidth penalty degrades the performance of LT-TG and along with transceiver penalty is responsible for LT-TG performing poorly.

4.3.3.3 Multi-Hop Summary

For single-hop mesh networks, LT performs the best in terms of blocking performance. For multi-hop transceiver constrained mesh networks, LP-TG is a compelling option. For multi-hop wavelength constrained mesh networks, SLT-TG is a good choice. SLT-TG has performance similar to LT-TG but does not need distributed arbitration and hence is preferred due to its simplicity of control. Bandwidth penalty and transceiver penalty are the price to be paid apart from additional hardware and software for using traffic aggregation techniques.

CHAPTER 5.   Survivable Network Design

Survivable network design has been a well researched topic in the area of LP and LP-TG networks [80]. Network survivability refers to the ability of the network to recover from node, link or equipment failures. Network resilience is vital considering the implications of failure on a WDM network of vast transport and switching capacities. The failure due to a fiber cut is not unusual and hence we study survivability mechanisms in the context of light-trail networks.

The original service route of a connection is called the primary trail. When a failure occurs on the primary trail, the connections on the trail get re-routed over backup trails. The survivability techniques can be broadly classified into two: protection and restoration. Protection refers to designing networks with spare capacities (wavelengths) so as to tolerate specific kinds of failure. For restoration, spare capacities are not provisioned in advance. In the event of a failure, an online mechanism is invoked that searches for spare capacities in the network through which the backup trail can be routed. Restoration is more efficient than protection in terms of resource utilization but is complex. Protection is better in terms of the ability to provide service guarantees.

Protection and restoration techniques can be further classified as link-based or connection-based in light-trail networks. In link-based protection, the connections are rerouted around the failed link. Nodes in the detour route let this trail bypass their LAUs. As observed in lightpaths [80], link-based protection is unattractive for trails as well, since the backup trails are be usually longer and the choice of backup trails is limited. Due to the absence of wavelength conversion, the wavelength continuity constraint should be adhered to which may lead to significant capacity overhead. Besides, handling node failures is not possible with the link-based approach.

In a connection-based approach, a backup trail is assigned to every primary connection. The backup capacity can be shared or dedicated. In the dedicated protection scheme (DP), data can be assumed to be sent on the primary and backup trails simultaneously. The destination uses the signal that has better quality. In the shared protection scheme (SP), the backup capacity is shared among multiple backup connections whose primaries do not fail together. As opposed to dedicated scheme, the backup capacities are reserved and used only in the case of a failure. DP requires more spare capacity than SP but is fast since the destination has to simply switch to the backup trail in the event of a failure. SP requires a complex protection switching mechanism that configures the crossconnects on the intermediate nodes to route the connection on the backup trail in the case of a failure.

If a link on a trail is cut, only connections traversing the cut gets affected. For instance, in a trail {0,1,2} serving connections (0,1) and (0,2), only connection (0,2) is affected when link (1,2) fails. The protection mechanism could depend on the fault location or could be independent of it. In a failure dependent scenario, only the affected connections on a trail are rerouted and the backup route can use all the links except the one that failed. In a failure independent scenario, all the connections on the trail get rerouted and the rerouted connections cannot traverse any of the links traversed by the primary trail. While fault dependent failure is more resource effective, the fault independent failure is easier to design and has lesser information to be maintained. All connections between a node pair traversing a single primary trail could be routed on multiple backup trails or be restricted to just one single backup trail. The signaling mechanism for the latter is simpler due to its course granularity but requires more spare capacity. The former mechanism is more fine granular, flexible and resource-efficient.

We explain a simple illustrative example to see why the survivability problem is hard. Consider a trail $T_1 = \{0, 5, 4, 3\}$ in the network shown in Figure 5.1 that serves requests $\{(0,5), (5,4), (0,3), (5,3)\}$. Each request is OC-12 and the capacity of a trail is OC-48. One possible solution to this problem is as follows. The backup trails corresponding to this would be $T_2 = \{5, 0, 1, 2, 3\}$, $T_3 = \{0, 1, 5\}$ and $T_4 = \{5, 1, 2, 4\}$. T_2 protects connections (5,3) and (0,3)

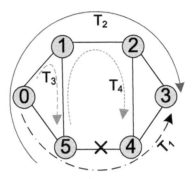

Figure 5.1 An example light-trail network with connection level protection

while T_3 and T_4 protect connections (0,5) and (5,4) respectively. The number of wavelength links required without protection is 3 while the number of wavelength links required with protection is 9. While there is two-fold increase in the number of wavelengths required, the trails T_2, T_3 and T_4 are significantly underutilized. The objective is to find edge-disjoint trails for every connection and at the same time maintain significant utilization on all the trails which is a challenging problem.

In this chapter, we first study dedicated and shared, connection-level, failure independent protection for static traffic. We design simple heuristics for survivable static design and evaluate its performance on random mesh networks. Next, we study shared, connection-level protection for dynamic traffic. We use the auxiliary graph approach introduced in Chapter 3 to develop a novel and simple approach for dealing with dynamic traffic. We evaluate its performance on random mesh networks.

5.1 Static Survivable Network Design

In this section, we investigate the costs (in terms of wavelengths) required to design a network that can survive all single link failures. Towards this end, we first present a formal description of the Static Trail Routing And Wavelength assignment (STRAW) problem and the involved assumptions. We formulate an ILP to solve the STRAW problem in Section 5.1.2. Since the ILP is computationally intractable for large scale networks, we design efficient heuristics that yield approximate solutions for both dedicated and shared segregated protection in Section 5.1.3. Our conclusion based on our simulation results shown in Section 5.1.4 is that with modest amount of spare capacity, 100 % single link failure recovery can be achieved.

5.1.1 Problem Definition

We make the following assumptions in our work. The network topology is 2-connected. All links are bi-directional. All nodes are equipped with tunable transceivers and wide-band receivers. Wavelength conversion capabilities are not present in the network. Individual connections between any node pair are sub-wavelength and is subject to non-bifurcation constraints. If multiple connections exist between a node pair, each one can be routed through a different path. We assume the presence of an out of band communication mechanism that allows all the nodes active in a trail to know about a failure that happens on any link traversed by this trail. With these assumptions, the problem can be stated as follows.

Given a directed network topology G(V,E), where V is the node set, E is the link set and given the traffic matrix R, design the network to optimize one of the objectives (1) Max throughput STRAW: Given the number of wavelengths on each link (W) and the number of transceivers on each node (X), identify the routing and wavelength assignment for primary and backup route so as to maximize network throughput.(b) Min cost STRAW: Identify the minimum network resources required to route and assign wavelengths for primary and backup route for every connection so as to survive all single link failures.

5.1.2 ILP Formulation

We formulate an integer linear program to solve the Min Cost STRAW problem. Dedicated protection scheme is assumed and the protection mechanism is failure independent. We describe our notation, give a formulation for maximizing STRAW throughput and modify it to solve the above two problems.

N - number of nodes in the network (data)

C - capacity of a wavelength (data)

W - number of wavelengths on each link of capacity C (data)

LT - set of possible light-trails in the network (data)

LT_t - an instance of a light-trail $LT_t \in LT$ (data)

LT_t^r - set of requests that can be supported by LT_t based only on the containment constraint (data)

$LT_t^{i,j}$ - 1 if trail LT_t traverses link (i,j), 0 otherwise (data)

$t, t_1, t_2 = 1..\|LT\|$ - number assigned to each light-trail(index)

I_{t_1,t_2} - 1 if t_1 and t_2 share at least one link, 0 otherwise (data)

$\lambda = 1..W$ - number assigned to each wavelength (index)

$i, j = 1..N$ - nodes in the network (index)

α - a very large number (say, 1000) (data)

$\Phi_{i,j}^{k,y}$ - 1 if k^{th} OC-y from i to j is carried by the network, 0 otherwise (variable)

$P_{i,j,t}^{k,y}$ - 1 if k^{th} OC-y from i to j is carried by the network, 0 otherwise (variable)

$S_{i,j,t}^{k,y}$ - 1 if k^{th} OC-y from i to j is carried by the network, 0 otherwise (variable)

T_t^λ - 1 if wavelength λ is assigned to trail t, 0 otherwise (variable)

TX_t^i - 1 if node i on trail t needs a transmitter, 0 otherwise (variable)

RX_t^i - 1 if node i on trail t needs a receiver, 0 otherwise (variable)

U_λ - 1 if wavelength λ is used, 0 otherwise (variable)

N_λ - number of wavelengths used in the network (variable)

T_t - number of instance of trail LT_t (variable)

T_{max} - number of trails on the maximum loaded link (variable)

C_t - the cost of a transceiver

C_λ - cost of adding a wavelength to a network

$$\text{Minimize} C_\lambda N_\lambda + C_t N_t \tag{5.1}$$

Subject to constraints

Primary assignment constraint

$$\sum_{t}^{(i,j)\in LT_t^r} P_{i,j,t}^{k,y} = \Phi_{i,j}^{k,y} \quad \forall k, y, i, j \tag{5.2}$$

Backup assignment constraint

$$\sum_{t}^{(i,j)\in LT_t^r} S_{i,j,t}^{k,y} = \Phi_{i,j}^{k,y} \quad \forall k, y, i, j \tag{5.3}$$

Capacity Constraints

$$\sum_{(i,j)\in LT_t^r} \sum_{k,y} y P_{i,j,t}^{k,y} + y S_{i,j,t}^{k,y} \leq T_t\, C \quad \forall t \tag{5.4}$$

Primary-Backup Disjointness Constraint

$$(P_{i,j,t_1}^{k,y} + S_{i,j,t_2}^{k,y})(1 - I_{t_1,t_2}) \leq 1 \quad \forall k, y, i, j, t_1, t_2 \tag{5.5}$$

$$T_t \leq \sum_{(i,j)} \sum_{k,y} P_{i,j,t}^{k,y} + S_{i,j,t}^{k,y} \quad \forall t \tag{5.6}$$

Wavelength Assignment Constraints

$$\sum_{\lambda} T_t^{\lambda} = T_t \quad \forall t \tag{5.7}$$

$$\sum_{t} T_t^{\lambda} \leq 1 \quad \forall \lambda, \{t : LT_t^{p,q} = 1, \forall (p,q) \in E\} \tag{5.8}$$

Transmitter Usage Constraints

$$TX_t^i \geq P_{i,j,t}^{k,y} + S_{i,j,t}^{k,y} \quad \forall k, y, t, i, \forall (i,j) \in LT_t^r \tag{5.9}$$

$$\sum_{t} TX_t^i \leq N_t \quad \forall i \in LT_t \tag{5.10}$$

Receiver Usage Constraints

$$RX_t^i \geq P_{i,j,t}^{k,y} + S_{i,j,t}^{k,y} \quad \forall k,y,t,i, \forall (j,i) \in LT_t^r \tag{5.11}$$

$$\sum_t RX_t^i \leq N_t \quad \forall i \in LT_t \tag{5.12}$$

$$U_\lambda \geq \sum_t T_t^\lambda / \alpha \quad \forall \lambda \tag{5.13}$$

$$N_\lambda \geq \lambda\, U_\lambda \quad \forall \lambda \tag{5.14}$$

$$P_{i,j,t}^{k,y},\ S_{i,j,t}^{k,y},\ \Phi_{i,j}^{k,y},\ T_t^\lambda,\ TX_t^i,\ RX_t^i, U_\lambda \in (0,1)$$

$$N_\lambda\, T_t \in I$$

Equation (5.1) maximizes the throughput. Equation (5.2) and (5.3) reserve a primary and backup trail for every connection that is accepted in the network. Equation (5.4) ensures that the sum of the capacities of all the primary and backup connections in a trail do not exceed the capacity of the trail. Equation (5.5) prevents the primary and secondary trail from sharing any link and equation (5.6) discards any trail that does not carry any request. Equation (5.7) allocates a wavelength for every trail and equation (5.8) prevents any two trails sharing the same edge from being assigned the same wavelength. Equations (5.9) and (5.11) count the number of transceiver equipments required on each node. A communication equipment is used on a trail only if a primary or a backup connection to or from the node is served by that trail. Equations (5.10) and (5.12) ensure that the number of communication equipments required do not exceed what is provisioned. While equation (5.13) keeps track of the wavelengths that are used, equation (5.14) identifies the wavelength of maximum index that has been already assigned.

(1) Min Cost STRAW: If we set $\Phi_{i,j}^{k,y} = 1 \ \forall k,y,i,j$ as an additional constraint, the above formulation models min cost STRAW problem. (2)Max Throughput TRAW: If all occurrences of $S_{i,j,t}^{k,y}$ are removed from the above formulation and the objective function **Maximize** $\sum_{i,j} \sum_{k,y} \Phi_{i,j}^{k,y}$

is used, and equations (5.3) and (5.5) are discarded, the resulting formulation models max throughput TRAW problem.

5.1.3 STRAW Heuristics

Heuristics for protection in large-scale lightpath networks was studied in [132, 129]. We propose two heuristics to solve the min cost problem in large-scale survivable light-trail networks. For our heuristics, we decouple the routing and wavelength assignment problems. Though this may lead to discrepancy between the approximate results and the optimal solution, our heuristic is simple and fast while an attempt to solve the combined problem becomes intractable for large networks. The output of our heuristic is a primary trail and a backup trail for every connection and the wavelengths assigned to them.

We first describe our routing and coloring heuristic which are common to both the protection schemes. For wavelength assignment, an auxiliary graph is obtained based on the results from the routing step. The rules for generating the auxiliary graph are different for dedicated and shared protection but the coloring heuristic used is the same. We describe the routing and coloring heuristic and the auxiliary graph generation rules.

5.1.3.1 Routing

We study connection-based failure independent protection and hence, for every connection, a primary route and a secondary route that is link disjoint with the primary route is identified. To find link disjoint routes, the heuristic method suggested in [132] is used. We define the term backup containment set for a trail. A backup connection (v_i, v_j) is said to be in the backup containment set of trail t, if node v_j is downstream of node v_i on trail t and if the trail t is edge disjoint with the primary trail of this connection.

Candidate backup trail packing is done in the following way. We list all the connections in the backup containment set of the trail in increasing order of request sizes. We pack all the elements from the left until taking one more would defy the capacity constraints of the backup

R	W	T
194	4	5
406	6	6
751	8	7
925	8	11

Figure 5.2   ILP results for min cost STRAW formulation on the six node network.

trail. Next, we assign a wavelength and transceivers to this trail on a first fit basis. We now explain the procedure for primary and backup routing.

We list the shortest routes between all node pairs with nonzero request in list P, which corresponds to the candidate primary trails listed in Step 1 of Table I. For every candidate primary route in P, we list the corresponding candidate edge disjoint secondary route in list S. By giving P as input to one of the heuristics in section IV, we assign a primary trail to every connection and place the primary trails in list Γ_p.

We pack all the candidate backup trails in S as explained above and choose the best backup trail and place in Γ_b. The best backup trail is the one with the highest packing fraction. Wavelength index and range are other metrics used to break possible ties. S is packed again and the next best candidate is selected. This step is repeated until each connection is assigned a backup trail. The backup trails are placed in list Γ_b.

5.1.3.2   Wavelength Assignment

After all primaries and secondaries are routed on the graph, each trail needs to be assigned a wavelength according to the wavelength assignment and continuity constraints. We construct an auxiliary graph, G', such that each light-trail in the system is represented by a node in G'. An undirected edge between two nodes is introduced in G' if the trails are required to have different colors. The nodes in the auxiliary graph are colored using the standard largest-first algorithm.

5.1.3.3   Auxiliary graph generation

Dedicated Trail Protection (DP)

The auxiliary graph G' required for coloring is generated as follows. For every trail t $\in \Gamma_p$ and Γ_v, introduce a vertex in G'. An edge is introduced between two vertices if the corresponding two trails share at least one link in the network.

Shared Trail Protection (SP)

The auxiliary graph G' is generated as follows. Consider two primary trails t_1 and $t_2 \in \Gamma_p$. Suppose primary trail t_1 carries connections c_1^1 through c_m^1. Let the backup trails corresponding to each connection be b_1^1 through b_m^1 with not all of them necessarily distinct. Let primary trail t_2 carry connections c_1^2 through c_n^2. Let their corresponding backup trails be b_1^2 through b_n^2 with not all of them necessarily distinct. For every trail obtained from the routing step, introduce a vertex in G'. Edges are introduced in graph G' based on the rules defined below.

1. t_1 is disjoint with b_1^1 through b_m^1, by definition. Hence, there is no edge between t_1 and any of b_1^1 through b_m^1. The same applies for t_2 and its backup trails.

2. If t_1 fails, all connections on t_1 are routed on its corresponding backup trails. Consider b_1^1 through b_m^1 pairwise. If the individual trails in a chosen pair are distinct and if they share a common link, introduce an edge between them. The same applies for the relation between t_2 and its backup trails.

3. If t_1 and t_2 are disjoint, there exists no edge between t_1 and t_2. An edge is not introduced between trails α and β, $\alpha \in \{b_1^1, .., b_m^1\}$, $\beta \in \{b_1^2, .., b_n^2\}$ in this step even if α and β share a link. This is because, t_1 and t_2 will not fail simultaneously and hence their backup trails will not be activated simultaneously.

4. If t_1 and t_2 overlap on a link, then introduce an edge between them. An edge is introduced between trails α and β, $\alpha \in \{b_1^1, .., b_m^1\}$, $\beta \in \{b_1^2, .., b_n^2\}$ if α and β share a link.

5. If t_1 overlaps with a backup trail of t_2, i.e, $\beta \in \{b_1^2, .., b_n^2\}$, introduce an edge between t_1 and β. Similarly if t_2 overlaps with a backup trail of t_1, introduce an edge between them.

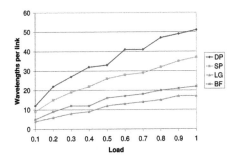

Figure 5.3   Wavelength requirement comparison among networks with no
protection, dedicated protection and shared protection for a
maximum aggregate traffic of 14,571 OC-1 units.

5.1.4   Simulation Results

We discuss results obtained from ILP for our min cost formulation. CPLEX 8.1.0 was
used to solve the integer program for the network shown in Figure 4.1(a). The capacity of a
wavelength is 48 units. The requests between a node pair are of three granularities - 1, 3 and
12 and are uniformly distributed in the range (0,1), (0,1) and (2,3) respectively leading to a
total offered traffic of 925 units.

In Figure 4.5, T refers to the number of tunable transceivers per node, W, the number
of wavelengths per link, and R, the total offered traffic. The results refer to the minimum
cost resource requirements for survivable network design under dedicated, connection-based,
failure-independent protection scheme. While the network with no protection needs $T = 3$
and $W = 5$ to accept all the traffic, the survivable network design requires $T = 11$ and $W =$
8, showing a significant increase in the redundancy requirement. When the offered traffic is
reduced, the resource requirement is also reduced as expected.

Next, we present the STRAW heuristic results for the 25 node test network shown in Figure
4.1(b). The granularities 1, 3 and 12 are uniformly distributed in the range (0,12), (0,4) and
(0,2) respectively. Figure 5.3 shows the number of wavelengths required as a function of load
(as defined above) for networks with and without protection when transceivers were provided

in plenty. As expected, networks without protection require the least amount of resources. Shared protection outperforms dedicated protection significantly. When the aggregate traffic in the network is about 14,571 units, the maximum number of wavelengths required by BF heuristic described in Section 3.3.3.2 is 22, while that for shared protection, it is 37. For dedicated protection, 51 wavelengths are required. A redundancy of about 100 % is required for shared protection and about 200 % is required for dedicated protection.

5.2 Dynamic Survivable Network Design

In this section, we develop a solution for survivable network design in the presence of dynamic traffic. In [84], three approaches for shared protection in the context of dynamic traffic in LP-TG networks were proposed - protection at the lightpath level, mixed protection at the connection level and separate protection at connection level. The trade-offs made between transceiver usage and wavelength usage were analyzed and the authors conclude that protection at the connection level performs better than protection at the circuit level for lightpath based networks. Two approaches for dedicated protection in LP-TG architecture were proposed and evaluated in [85].

In the discussion that follows, we present some of the assumptions made in our work, the constraints related to the problem, the objective we want to accomplish and a description of the algorithm we implement followed by a discussion on the simulation results we obtained.

5.2.1 Assumptions

- The physical topology is given by $G = (V,E)$, where V is the set of nodes and E is the set of fiber links connecting the nodes.

- The capacity of each wavelength is arbitrarily assigned to be $C = 10$, the number of wavelengths on each link is W and the number of transceivers on each node is X. The transceivers are tunable to any wavelength in the fiber.

- The call arrivals to a network are assumed to be poisson distributed and the call depar-

Input : $G(V,E)$, $< s, d, B, t_h >$, $C_o(e)$, C, ϕ_e^t and all other information for every circuit t, wavelength state and transceiver state, where s - source, d - destination, t_h - holding time, B - connection bandwidth, $C_o(e)$ - cost of edge e, and C - capacity of a wavelength.

Output: Link disjoint working and backup circuit.

STEP 1: Compute the auxiliary graph based on the wavelength state, transceiver state and existing circuit information.

STEP 2: Prune edges that cannot handle the request capacity B. Compute minimal cost path from the output vertex of the access layer on node s to the input vertex of the access layer on node d to be p_w.

STEP 3: If primary is not routable, drop the call. Otherwise, proceed to the next step.

STEP 4: Compute minimal cost path p_b from the output vertex of the access layer on node s to the input vertex of the access layer on node d by assigning a new cost $C'(e)$ to every edge e as follows:

    a. if edge e does not represent a link or a circuit, $C'(e) = C_o(e)$

    b. if edge e represents a link, $C'(e)$ is

            – $+\infty$ if e is not route disjoint with p_w or link does not have a free wavelength

            – $BC_o(e)$ otherwise

    c. if edge e represents a circuit k, $C'(e)$ is

            – $+\infty$ if circuit k is not route disjoint with p_w or $(\phi_t^* - \phi_t^{e'})$ plus the residual capacity of k is less than B for some edge e' traversed by p_w

            – ϵ if circuit k is route disjoint with p_w and $(\phi_t^* - \phi_t^{e'})$ plus the residual capacity of k is no less than B for some edge e' traversed by p_w

            – $B'C_o(e)$ otherwise, where $B' = B - min\{\phi_t^* - \phi_t^{e'}\}$ for every edge e' traversed by p_w

STEP 5: If secondary is not routable, drop call. Otherwise, proceed to the next step.

STEP 6. Update network state and circuit information corresponding to backup route p_b. For every link e traversed by p_w, $\phi_t^e = \phi_t^e + B$. Compute the new value of ϕ_t^*. If required, remove wavelength links and add/modify edges in the virtual topology layer.

Figure 5.4 Auxiliary graph based traffic grooming heuristic

tures are assumed to be exponentially distributed with a rate normalized to unity.

- The calls are uniformly distributed between arbitrary source destination pairs.

- The call sizes are sub wavelength in nature and individual granularities have equal capacity arrival rates.

5.2.2 Constraints

Wavelength Continuity Constraints: We assume that wavelength conversion is not present in the network. So, there is a requirement for a wavelength continuous route from source to destination.

Resource Constraints: The number of wavelengths per link and number of transceivers per node are limited and pre provisioned.

Diverse Routing Constraints: The primary path should be link wise disjoint from the secondary path.

Circuit Capacity Constraints: A wavelength in a network consists of the following four dynamic partitions:

- Primary capacity

- Dedicated backup capacity

- Shared backup capacity

- Residual capacity

The sum of the sizes of these partitions should add up to the capacity of a wavelength.

5.2.3 Objective

The objective is to minimize blocking performance and identify a path disjoint backup route for every primary route. This can be achieved by maximizing backup bandwidth sharing without violating capacity constraints.

5.2.4 Algorithm

In our current work, we exploit the concept of backup sharing to minimize blocking performance. Shared backup leads to capacity savings at the expense of reduced restoration speeds and can be explained as follows. Assume a circuit t of capacity 10 units packed with 6 units of primary traffic. The 4 units spare capacity can carry a backup connection c of size 4 units if c is in the backup containment set of t. This spare capacity can also be shared as well. If two backup connections of size 4 units each conform to the backup containment set of t, then both could share the 4 units spare capacity on t, if their corresponding primaries will not fail together. To keep track of the shared backup capacity, the following information is maintained for every trail t:

- The route, the wavelength assignment and the residual capacity of trail t

- For every link e, ϕ_t^e identifies the amount of traffic to be routed over the trail t if link e fails.

- ϕ_t^* refers to the maximum value of ϕ_e^t for all e in the network.

The parameter ϕ_t^e captures information related to backup sharing. ϕ_t^* is the amount of backup capacity reserved on circuit t. The value of $(\phi_t^*-\phi_t^e)$ refers to the amount of free bandwidth available on circuit t for backing up a connection that traverses link e. Using this free bandwidth available on circuits and the residual capacity on links and circuits, the backups are routed so as to minimize the used extra capacity. The details of our algorithm are provided in Figure 5.4 and described below.

When a call arrives between source s and destination d, we compute the auxiliary graph based on information related to existing circuits, wavelength and transceiver state, cost of the edges and the backup sharing information ϕ_t^e for every edge e. Based on dijkstra's algorithm, the shortest route is found from the output port of the access layer of node s to the input port of the access layer of node d. If the primary is not routable, the call is dropped. The transceiver and wavelength status are updated temporarily so as to find the backup route. An edge disjoint backup route from s to d is computed after assigning new cost to all the edges

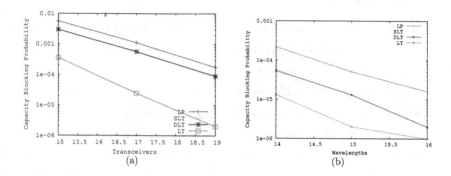

Figure 5.5   (a) Throughput as a function of number of transceivers per node
for X = 25 and R = 125 E (b) Throughput as a function of W
= 23 and R = 50 E

in the network. It is possible to route the backup connection of a call on a circuit edge such
that the additional bandwidth consumed is either minimal or zero. The costs of the edges are
made proportional to the amount of additional capacity required by the edge to support the
backup connection. By selecting the minimal cost route, the additional bandwidth required
to support this connection is minimized at the network level. If there exists no backup route,
the call is dropped. Otherwise, the information for backup multiplexing is updated on all the
trails that carry the backup connection. The call is accepted and all the required wavelength
and circuit information are updated.

5.2.5   Simulation Results

We implemented the dynamic protection model discussed above using discrete event simulation techniques. The simulations were performed on random topologies that were generated
based on the Waxman model discussed in Chapter 4. The parameters, $\alpha = 0.4$, and $\beta = 0.3$
were used to generate a random two connected mesh network with 20 nodes, 73 links, and a
diameter of 5. The simulations were done for both scenarios - transceiver constrained scenario
and wavelength constrained scenario. A million calls were generated and capacity blocking

performance is reported after a steady state was achieved.

The wavelength constrained scenario is reported in Figure 5.5(a) with $X = 25$ and Load $= 125$ Erlangs. The transceiver constrained scenario is reported in Figure 5.5(b) with $W = 23$ and Load $= 50$ Erlangs. From both the figures, it is clear that LT performs the best, closely followed by SLT and DLT and LP performs the worst. Naturally, when the transceivers or wavelengths are increased, blocking reduces. There is more than an order of magnitude difference between LT and LP in terms of capacity blocking performance.

5.3 Conclusions

In this chapter, we discussed possible protection mechanisms in light-trail networks, reasoned why STRAW is a challenging problem and designed heuristics for dedicated and shared connection level segregated protection. We found that, with dedicated protection, about 200 % redundancy may be required. Shared protection performs much better and full protection can be achieved in the presence of single link failures with less than 100 % redundancy. Our current static survivability work considers segregated protection.

We described an auxiliary graph based algorithm for survivable network design with dynamic traffic. We used this idea to compare the performance of LP, SLT, DLT and LT architectures. We observed that for random mesh topologies with transceiver and wavelength constrained scenarios, LT performs orders of magnitude better than LP architecture.

Our ILP results for static traffic suggest that optimal solutions involve trails that multiplex both primary and secondary connections. We would like to design heuristics for mixed protection strategy in the future. We also intend to consider multi-hop scenarios and compare the performance of different architectures.

CHAPTER 6.   A MAC protocol for light-trail networks

A number of priority based access protocols have been proposed for shared wavelength networks in the past. Two issues need to be considered in the context of unidirectional networks. First, there may not exist a circuit in the reverse direction. Second, even if there exists a circuit in the reverse direction, the ratio of propagation delay to packet time may be high in optical networks leading to a drop in capacity if feedbacks are required on a per packet basis. In [82],a protocol called pi-persistent protocol is proposed for multi-access communication on unidirectional time slotted bus networks with fixed packet sizes. In the pi-persistent protocol, node i persists in transmitting into an empty slot with a probability pi. The pi values can be optimized to achieve equal-throughput, equal-delay, equal-blocking, or any ratio of throughput shares between the nodes. This mechanism is extended for variable length packets in [81] by fragmenting them into many fixed-size packets and using the bit-continuation approach. A scheme based on probabilistic scheduling strategy is discussed in [10], and algorithms for scheduling variable-length message transmissions with a single control packet and using a separate control channel are presented in [64]. The following sections describe two medium access control methods used in light-trail networks.

6.1   The Light-trail Protocol

A simple MAC protocol based on carrier sensing is proposed for LTs in [52]. When a node, say X, wants to transmit data, it senses the carrier for upstream activity. If the channel is free, X sends a beacon signal downstream indicating that it has a packet to transmit. If the channel is busy, X waits until the channel is free to send the beacon signal. After some predetermined offset time, called the guard band, X transmits its data. Node X continues

159

sensing the channel while its packet is being transmitted. During transmission, X may hear a beacon signal from an upstream node, say Y, which has data to transmit. Upon hearing this, X terminates its transmission and lets Y's packet pass through. X's truncated packet is discarded by the receivers since it fails the link-level error checks. Thus, by sending a beacon signal before transmitting a packet and always giving priority to the upstream nodes, the protocol successfully resolves medium access contentions. The key design parameter that affects protocol performance is the guard band gap. When the decision to stop is made, feedback control happens through a microcontroller; hence, the delay is large (typically at least 50 ns) due to the electronic processing overhead. The guard band (set to 75 ns) should be set large enough to pull out of transmission after sensing the beacon signal and small enough to avoid significant overhead.

6.2 The Lightbus Protocol

The lightbus medium access control protocol works as follows. In the lightbus architecture, a fiber delay loop is included on every node for every signal that is coming into the node. The delay accounts for the time needed to transmit a maximum size packet. This delay time allows the transmitter to complete its current transmission when a transmission from an upstream node is detected on the lightbus.

When a node has data to transmit, it first checks for activity on the delay line. If the delay line is free, the node goes ahead and completes its transmission. If the delay line is busy, the node waits till the delay line becomes free and then starts transmission. In the middle of a transmission, if a node upstream starts sending data, it is buffered in the delay line of the downstream node until the current transmission on the node is completed.

The key idea in the lightbus architecture is to design a more amenable and efficient control structure on both the sender and the receiver nodes. The transmitter does not have to worry about dropping, suspending and retransmitting packets and the receiver does not have to deal with runt packets. Unlike the light-trail protocol, the lightbus approach allows for all the nodes in the bus to send out data simultaneously (under certain conditions) and does not lead to wasted transmissions that result in increased power consumption. Besides, a guard band on every packet is required by the light-trail protocol whose duration depends on the rise and fall times of the transponders and the speed of the control electronics. This additional overhead incurred for every packet is avoided in the lightbus protocol. However, the presence of the fiber loops impose extra queueing and propagation delays. Thus, the central idea behind the lightbus model is to trade-off performance for simplicity and efficient control.

The lightbus and light-trail medium access protocols are modeled in section 6.3 and the simulation results are analyzed in section 6.4. Section 6.5 presents a discussion on observed behavior of the protocols based on the analytical model and suggests some performance enhancements to the lightbus protocol. Finally, we present some conclusions based on our current research and outline future directions.

6.3   Analytical Model

In this section, we present the exact model of the delay characteristics of a 3-node Light-trail and the 3-node Light Bus protocols. It becomes difficult to extend the model to more than three nodes as the traffic characteristics changes after the second node. However, an approximate model that aggregates the traffic coming from upstream nodes and representing it by a poisson process is feasible and will be identical to the model presented in this section. To facilitate our presentation, the following notations will be used for packets on each queue. Note that the definitions of some of these terms are explained in the next subsection.

W_2 - Virtual service time of a node 2 packet that arrives to a non-empty queue.

W_2^0 - Exceptional service time of a node 2 packet that arrives to an empty queue.

C_2 - Completion time of a node 2 packet.

B_i - Duration of the busy period.

E - Duration of the extended busy period.

S_i - Duration of silence.

R_E - Residual life time of E seen by an incoming packet.

$\phi_E(z)$ - Moment Generating Function (MGF) of E.

$\phi_S(z)$ - MGF of S.

$\phi_{R_E}(z)$ - MGF of R_E.

$\phi_T(z)$ - MGF of waiting and service time of a node 2 packet.

$\bar{B}, \bar{E}, \bar{R}_E$ - Mean values of the busy period, extended busy period and the residue of the extended busy period.

b_1, b_2 - Service time of a node 1 and node 2 packet.

$\bar{b_1}, \bar{b_2}$ - Mean values of node 1 and node 2 packet service time.

λ_1, λ_2 - Packet arrival rates at node 1 and node 2.

ρ_1, ρ_2 - Utilization factor of queues on node 1 and node 2.

D - Maximum service time of a packet.

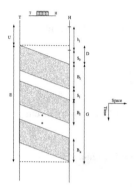

Figure 6.1   Space-Time diagram for the Light Bus model. The two vertical
lines T and H refer to the two end points of the delay line on
node 2. T denotes the tail of the delay line where the packet
enters and H denotes the head of the delay line where the packet
exits.

6.3.1   The lightbus model

We analyze the delay distribution characteristics of a three-node Light Bus which starts
from node 1, passes through node 2 and ends at node 3. The third node of the Light Bus is not
involved in transmission of data and hence only the first two nodes are considered. We ignore
the propagation delay since the Poisson arrival process is stationary and hence unaffected by
time shifts.

6.3.1.1   Model for the first node

The behavior of queue 1 can be modeled as an M/G/1 system since the service time of the
packets of queue1 is independent of the arrivals on other queues. The Laplace transform of
the waiting time PDF is given by the Pollaczek-Khinchin transform equation [68] as

$$\phi_{W_1}(z) = \frac{z(1-\rho_1)}{z - \lambda_1 + \lambda_1 \phi_{B_2}(z)} \tag{6.1}$$

The mean queuing delay of the packets on node 1 is given by

$$W_1 = -\phi'_{W_1}(0) = \lambda_1 \bar{b}_1^2 / 2(1 - \rho_1) \tag{6.2}$$

6.3.1.2 Model for the second node

To understand the behavior of the second queue, refer to the space-time diagram in Figure 6.1. The space-time diagram depicts a temporal sequence of transmission activity at the tail end and the head end of the delay line on node 2. A transmission of a node 1 packet marks the beginning of a busy period, that is defined to begin with the arrival of a packet to an idle channel and to end when the channel next becomes idle. During this busy period, node 2 cannot transmit any of its packets. Consider the last packet transmitted by node 1 during such a busy period. By definition of the busy period, there is no other arrival on node 1 until this packet is completely serviced. In the time it takes for the trailer of this packet to clear the head end of the delay line (which is D time units), if there is another arrival on node 1, the second busy period starts and a packet from queue 2 (if any) still cannot enter service. This cycle repeats itself until the time when the last packet of a busy period clears the head end of the delay line without a new arrival starting off another busy period on node 1. We refer to the time from the beginning of the first busy period, as seen from the head end of the delay line, until the end of last busy period, as the "extended" busy period. The extended busy period, denoted by E_2 in Figure 6.1, is marked by a series of active periods (denoted by B_2^i) interspersed with periods of silence (denoted by $S_2^i, 0 < S_2^i < D$). We also define another random variable G_2, which is the time from the beginning of the first busy period, as seen from the Tail end of the delay line, until the end of last busy period, and is related to E_2 as,

$$E_2 = G_2 + D$$

Queue 2 can be considered to be a variant of an M/G/1 system where the first message of each busy period (i.e., a node 2 packet that comes to an empty queue) sees an exceptional virtual service time of W_2^0 and the other packets see a virtual service time of W_2 as explained below.

Consider a packet that arrives to an empty queue at node 2. It is a well known property that Poisson Arrivals See Time Averages (PASTA). So, with a probability q_2 (to be derived later), the packet finds the transmission line inactive and it immediately enters service. With a probability $1 - q_2$, the transmission line is active and the packet waits for remaining time of E, called X_2, before it is serviced. Irrespective of this, the service time of the queue 2 packet is denoted by the random variable b_2. It is possible that during b_2, a busy period B_2^1 starts at queue 1, which is seen at the head end of the delay line after a silence period of S_2^0, $S_2^0 < D$. The period between the start of the service time b_2 and the start of busy period B_2^1 (as seen at the tail end of the delay line) is denoted by U_2. It is to be noted that $U_2 + D = b_2 + S_2^0$. The first busy period B_2^1 may be followed by a series of B_2^i and S_2^i periods. Therefore, the total time elapsed after head of the queue 2 gets serviced and before the next packet (if any) on queue 2 can go into service is given by $S_2^0 + G_2$ as seen in Figure 6.1, where

$$G_2 = \sum_{i=1}^{k-1} (S_2^i + B_2^i) + B_2^k$$

and k is the last busy period, as seen on the head end of the delay line, during which an arrival does not happen on queue 1.

The exceptional virtual service time of the first packet is therefore given by,

$$W_2^* = X_2 + b_2 + S_2^0 + G_2 \tag{6.3}$$

The virtual service time of the other packets begin when the corresponding queue 2 packet begins transmission and ends when the packet is serviced and the system is clear of all queue 1 packets. So, the virtual service time is given by

$$W_2 = b_2 + S_2^0 + G_2 \tag{6.4}$$

The MGF of time spent by a node 2 packet in the system (including the queueing time and the service time), $\phi_{T_2}(z)$, can be obtained if $\phi_{W_2}(z)$ and $\phi_{W_2^*}(z)$ are known and is given by [107] to be,

$$\phi_{T_2}(z) = \frac{[1-\rho_2][\lambda_2\phi_{W_2}(z)-(\lambda_2-z)\phi_{W_2^*}(z)]}{[1+\lambda_2\overline{W_2^*}-\rho_2][z-\lambda_2+\lambda_2\phi_{W_2}(z)]} \tag{6.5}$$

The mean value of the time spent by a customer in the system, $\overline{T_2}$, can be found by inverting $\phi_{T_2}(z)$ to be ,

$$\overline{T_2} = \frac{\overline{W_2^*}}{1-\lambda_2\overline{W_2}+\lambda_2\overline{W_2^*}} + \frac{\lambda_2\overline{W_2^2}}{2(1-\lambda_2\overline{W_2})} + \frac{\lambda_2(\overline{W_2^{*2}}-\overline{W_2^2})}{2(1-\lambda_2\overline{W_2}+\lambda_2\overline{W_2^*})} \tag{6.6}$$

where $\overline{W_2^*}$ is the mean exceptional virtual service time and $\overline{W_2}$ is the mean virtual service time of a node 2 packet.

The mean queueing delay of the node 2 packet, $\overline{Q_2}$, can be found to be

$$\overline{Q_2} = \frac{\lambda_2\overline{W_2^2}}{2(1-\lambda_2\overline{W_2})} + \frac{\lambda_2(\overline{W_2^{*2}}-\overline{W_2^2})}{2(1-\lambda_2\overline{W_2}+\lambda_2\overline{W_2^*})} \tag{6.7}$$

The derivations of expressions for $\phi_{B_2}(z), \phi_{b_2+S_2^0}(z), \phi_{S_2^i}(z), \phi_{W_2}, \phi_{X_2}(z), \phi_{W_2^*}$ in that order are presented below.

From [68], the expression for $\phi_{B_2}(z)$ is given by,

$$\phi_{B_2}(z) = \phi_{b1}[z+\lambda_1-\lambda_1\phi_{B_2}(z)]$$

Next, we find an expression for the joint density function of $b_2+S_2^0$. b_2 and S_2^0 are dependent random variables. We have

$$f_{b_2+S_2^0}(b_2+S_2^0 = t+v| \text{ an arrival happens at queue 1 in time } b_2 = t)$$

$$= \frac{\lambda_1 e^{-\lambda_1(t+v-D)}}{1-e^{-\lambda_1 t}}$$

since $u+D = t+v$ as seen in Figure 6.1.

$$\phi_{b_2+S_2^0}(z) = \int_{t=0}^{t=D}\int_{v=D-t}^{D} \frac{\lambda_1 e^{-\lambda_1(t+v-D)}e^{-(t+v)z}(1-e^{-\lambda_1 t})}{(1-e^{-\lambda_1 t})} f_{b_2}(t)dvdt$$

After some manipulations, we get

$$\phi_{b_2+s0}(z) = \frac{\lambda_1 e^{-zD}}{\lambda_1+z}[1-\phi_{b_2}(\lambda_1+z)] \tag{6.8}$$

For the silence period, $S_2^i, i \geq 1$, we have,

$$\phi_{S_2}(z) = \int_0^D \frac{\lambda_1 e^{-\lambda_1 t} e^{-zt}}{1 - e^{-\lambda_1 D}} dt$$

$$= \frac{\lambda_1}{\lambda_1 + z} \frac{1 - e^{-(\lambda_1 + z)D}}{1 - e^{-\lambda_1 D}} \tag{6.9}$$

The virtual service time W_2 is simply the service time of the packet on node 2 if no arrival happens at node 1 until the packet has completed service. However, if an arrival happens during the period, an extended busy period is triggered as discussed above.

$$\phi_{W_2} = \phi_{b_2}(\lambda_1 + z) + \phi_{b_2 + S_2^0}(z)\phi_{B_2}(z)e^{-\lambda_1 D}\sum_{i=1}^{\infty}[\phi_{B_2^i}(z)\phi_{S_2^i}(z)]^{(i-1)} \tag{6.10}$$

since B_2^i are i.i.d and S_2^i are i.i.d for all $i \geq 1$.

$$\phi_{W_2}(z) = \phi_{b_2}(\lambda_1 + z) + \frac{\phi_{b_2 + S_2^0}(z)\phi_{B_2}(z)e^{-\lambda_1 D}}{1 - \phi_{B_2}(z)\phi_{S_2}(z)} \tag{6.11}$$

where the expression for $\phi_{b_2}(\lambda_1 + z)$ is given by

$$\phi_{b_2}(\lambda_1 + z) = \int_{t=0}^{t=\infty} e^{-(\lambda_1 + z)t} f_{b_2}(t)dt \tag{6.12}$$

$$\phi_{W_2^*} = \phi_{W_2}\phi_X(z) \tag{6.13}$$

where $\phi_X(z)$ can be derived as follows. Let p_1 be defined as the probability that a customer from queue 1 arrives before a customer on queue 2. Let p_2 be defined as the probability that a customer on queue 2 arrives before a customer on queue 1. Since, arrivals are exponential, the expression for p_1 and p_2 are as follows:

$$p_1 = \frac{\lambda_1}{\lambda_1 + \lambda_2}, p_2 = \frac{\lambda_2}{\lambda_1 + \lambda_2} \tag{6.14}$$

Consider the instance when both the queues are empty. Now, several extended busy periods of E_2 may be completed and no packet may arrive on queue 2 yet. When finally, a packet arrives on queue2, it may come before the arrival of the first packet to an empty queue1, which happens with probability p_2. Alternatively, it may arrive after the first packet arrived to the empty queue1 starting an extended busy period and it may have to wait for the residual time of the extended period. Hence,

$$\phi_X(z) = \sum_{n=0}^{\infty} [p_1\phi_{E_2}(\lambda_2)]^n p_2 \int_{r=0}^{\infty} e^{-zr}\delta_r(0)dr \tag{6.15}$$

$$+ \sum_{n=0}^{\infty} [p_1\phi_{E_2}(\lambda_2)]^n p_1 \int_{t=0}^{\infty} \int_{r=0}^{t} \lambda_2 e^{(-\lambda_2(t-r))} e^{-zr} dr f_{E_2}(t)dt \tag{6.16}$$

$$\phi_X(z) = \frac{\lambda_2}{\lambda_1 + \lambda_2 - \lambda_1\phi_{E_2}(\lambda_2)} + \frac{\lambda_1\lambda_2[\phi_{E_2}(\lambda_2) - \phi_{E_2}(z)]}{[\lambda_1 + \lambda_2 - \lambda_1\phi_{E_2}(\lambda_2)][z - \lambda_2]} \tag{6.17}$$

Note that to solve for $\phi_X(z)$, expression for $\phi_{E_2}(z)$ and $\phi_{E_2}(\lambda_2)$ required which can be found as follows:

$$E_2 = D + B_2 + \sum_{i=0}^{n-1}(B_2^i + S_2^i) + B_2^n \tag{6.18}$$

with probability $p_2^n(1 - p_2)$ where the probability of at least one arrival in D time units is

$$p_2 = 1 - e^{-\lambda_1 D} \tag{6.19}$$

B_2^i and S_2^i are sequences of i.i.d random variables with MGFs $\phi_{B_2}(z)$ and $\phi_{S_2}(z)$ respectively. So, the MGF of E is

$$\phi_{E_2}(z) = \phi_D(z)\phi_{B_2}(z)\sum_{n=0}^{\infty} \phi_{B_2}(z)^n \phi_{S_2}^n(z) p_2^n(1 - p_2) \tag{6.20}$$

$$= \phi_D(z)\phi_{B_2}(z)\frac{1 - p_2}{1 - p_2\phi_{B_2}(z)\phi_{S_2}(z)}$$

Since D is a constant, we have

$$\phi_D(z) = e^{-zD} \tag{6.21}$$

The MGF of the busy period is

$$\phi_{B_2}(z) = \phi_{b1}[z + \lambda_1 - \lambda_1\phi_{B_2}(z)] \tag{6.22}$$

$$\phi_{E_2}(z) = \frac{e^{-(\lambda_1+z)D}\phi_{b_1}[z + \lambda_1 - \lambda_1\phi_{B_2}(z)]}{1 - \frac{\lambda_1}{\lambda_1+z}\phi_{b_1}[z + \lambda_1 - \lambda_1\phi_{B_2}(z)](1 - e^{-(\lambda_1+z)D})} \tag{6.23}$$

The mean value of E_2 can be obtained from the first derivative of the MGF of E_2 as

$$\bar{E}_2 = -\phi'_{E_2}(0) = \frac{e^{\lambda_1 D} - 1}{\lambda_1} + \frac{\bar{b_1}e^{\lambda_1 D}}{1 - \lambda_1\bar{b_1}} \tag{6.24}$$

$\phi_{E_2}(\lambda_2)$ can be found by substituting λ_2 for z as follows,

$$\phi_{E_2}(\lambda 2) = \frac{e^{-(\lambda_1+\lambda_2)D}\phi_{B_2}(\lambda_2)}{1 - \phi_{B_2}(\lambda_2)\phi_{S_2}(\lambda_2)} \tag{6.25}$$

6.3.2 The light-trail model

In the Light-trail approach, any upstream node can interrupt a downstream node and the downstream node retransmits the failed packet after the upstream node completes its transmission. The retransmission can also be interrupted and hence, the downstream node may have to try repeatedly until its packet is successfully transmitted. As in the case of a Light Bus, the first node can be treated as a simple M/G/1 queue and the Pollaczek-Khinchin transform equation can be applied to obtain the queueing delay of the node 1 packets.

The second node can be modeled as a preemptive priority queue with restart-identical discipline. This is a well known problem and a solution for this is discussed in [62]. The completion time C_2 of a queue 2 packet begins when the system is clear of queue 1 packets and ends when service time b_2 has been completed and the system is clear of queue 1 packets. Notice that C_2 includes all interruptions by queue 1 packets, as well as all the service time of the queue 2 packet which must be repeated. The beacon signal transmission time and guard band gap that are required by the Light-trail protocol are accounted for by adding this time to the service time of the packet. The first and the second moments of the completion times are found to be [62]

$$E(C_2) = [\frac{1}{\lambda_1} + E(b_1)]E[e^{\lambda_1 b_2} - 1] \tag{6.26}$$

$$E(C_2^2) = 2[\frac{1}{\lambda_1} + E(b_1)]^2 E[(e^{\lambda_1 b_2} - 1)^2][E(b_1^2) + \frac{2E(b_1)}{\lambda_1} + \frac{2}{\lambda_1^2}] \times \qquad (6.27)$$

$$[E(e^{\lambda_1 b_2}) - 1] - 2[E(b_1) + \frac{1}{\lambda_1}] \int_0^\infty x_2 e^{\lambda_1 x_2} b_2(x_2) dx_2 \qquad (6.28)$$

The average waiting time of the packets is given by [126]

$$\bar{W}_2 = \frac{\lambda_2 E(C_2^2)}{2[1 - \lambda_2 E[C_2]]} + \rho_1 \bar{R}_E \qquad (6.29)$$

It is to be noted that no product form expression for the MGF of C_2 can be obtained due to the fact that the restart-identical model of service implies a non-memory less process.

6.4 Simulation Results

Performance analysis of the protocols for different traffic loads are done using discrete event simulation techniques. Infinite buffers are assumed to be present on all the nodes and the arrival process is Poisson distributed. The simulations are done for 10 Gbps systems and the results are obtained after transmitting 10 million packets. We assume the guard band required by the Light-trail to be 75 ns. In all the graphs, the x-axis plots the total offered load expressed as a fraction of the capacity of the wavelength and the y-axis plots the delay normalized to the time taken to transmit a maximum sized packet. The traffic distribution is computed as follows.

Consider a Light Bus or Light-trail of length N. The convener node is Node 1 and the end node is Node N. The traffic is assumed to be uniform for all s-d pairs. If T is the total traffic carried by the Light Bus, t_i is the traffic between any two s-d pairs, and C is the total capacity of the lightbus, then the traffic sourced by a node i is given by

$$t_i = \frac{(N - i)T}{N(N - 1)/2} \qquad (6.30)$$

and the system has been designed such that $\sum_{i=1}^{N-1} t_i \leq C$

The packet sizes are uniformly distributed between 500 bytes and 1500 bytes. The traffic offered by nodes 1 and 2 are in the ratio 2:1 respectively. We see that the delay of the node

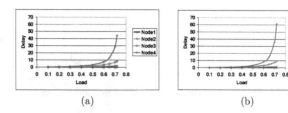

(a)   (b)

Figure 6.2   Queueing Delay Vs Load in (a) Light Bus (b) Light-trail when
packet sizes are uniformly distributed between 500 and 1500
bytes.

2 packets is very high and increases exponentially with increase in load since the node 1
packets are always given higher priority leading to starvation of node 2 packets. We performed
simulations for this three node case and found our simulation results (not shown here) to match
well with our analytical model.

We also performed simulations for bigger systems that are not covered by the model. We
considered five-node Light-trails and Light Buses since the study in [52] observes that the
average expected length of the Light-trails is five. The limit in the number of nodes occurs
because of power budget constraints. The input power should be large enough to allow tapping
of the signal in the intermediate nodes and yet small enough to prevent undesired non-linear
interactions in the fiber.

By Equation 6.30, for a five node system, traffic sourced by nodes 1 through 4 are in the
ratio 4:3:2:1 respectively. Our general observation is that both the Light-trail and the Light
Bus perform well until about loads of 0.7 C after which the delay increases exponentially. The
delay encountered by a packet on any node is less than the delay for a packet on nodes that are
further downstream. For results shown in Figure 6.2, the packet sizes are uniformly distributed
between 500 and 1500 bytes. When the packet sizes are small, the guard band required by the
Light-trail protocol is of the order of the packet transmission time and the Light Bus protocol
is able to outperform the Light-trail protocol by avoiding this overhead. In Figure 6.3, packet

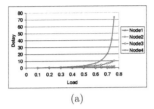 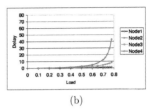

(a)   (b)

Figure 6.3   Queueing Delay Vs Load in (a) Light Bus (b) Light-trail when aggregated packet sizes are uniformly distributed between 16KB and 32 KB.

aggregation is assumed and the sizes are uniformly distributed between 16 KB and 32 KB. In this case, the guard band overhead is negligible when compared with the aggregate packet transmission time and hence the Light-trail protocol does better. It is observed that, in both the cases, the queueing delay is highest for the $N - 1^{th}$ node while the delays are negligible for all the other nodes. This is understandable since the upstream nodes are always given higher priority over the downstream nodes and the penultimate node suffers the most.

6.5   Discussion and Future Work

The Light-trail approach yields a slightly better performance than the Light Bus approach for bigger packet sizes since it transmits a packet at queue 2 at the end of the busy period from queue 1, which has a probability of success equal to $\phi_{b_2}(\lambda_1)$. Since the Light Bus approach always forces queue 2 packets to defer their transmission if they observe a packet coming downstream from queue 1 during D time units, then even short packets, which can be otherwise accommodated under the Light-trail approach, will have to wait under the Light Bus approach. That is, the probability of success under the Light Bus approach is $e^{-\lambda_1 D}$, which is always less than $\phi_{b_2}(\lambda_1)$ because $0 < b_2 \leq D$. We propose to improve upon the performance of the Light Bus protocol by implementing the following ideas.

The introduction of the delay line in the Light Bus approach actually gives an advantage, which is not present in the light-trail approach, that can be capitalized. Namely, we know what is going to happen during the next D time units. We therefore, propose an approach which will yield a throughput, which is as good as, or even better than that produced by the light-trail approach. When a node has data to transmit, if the delay line is empty, the packet at the head of the queue is transmitted. If the delay line is not empty, one of the following is done: if the gap on the delay line would allow the transmission of the packet at the head of queue 2, then it is transmitted. Otherwise, a fragment of the packet is transmitted, which can be easily done, and is supported by both IPv4 and IPv6. In order to reduce the fragmentation processing overhead and to streamline the fragmentation process, predefined fragment sizes can be used, like 1/4 MTU, 1/2 MTU and 3/4 MTU.

The processing overhead due to fragmentation depends on which protocol is being used. If the standard IPv4 protocol (or even IPv6) is used, then the overhead is basically computing the checksum and setting the values for the More, Offset and the Length fields. The checksum for all the bits in the header except for these fields are precomputed. Once the fragment size is decided, which can be 1/4, 1/2, 3/4 or 1 MTU, the offset field is set to this word, the More bit is updated and these bits are added to the checksum to generate the final checksum. Therefore, number of operations required for fragmentation are minimal and the introduction of fragments will not lead to a significant overhead. Also, unlike in a conventional packet switched network, packet reordering and packet losses are not serious issues in light-trail and Light Bus networks and hence fragmentation does not lead to performance degradation.

An alternate method to improve performance would be to consider the packets to be having the size of 1/4 MTU and allow burst mode transmission that can be of length 1, 2, 3 or 4 packets. This is a close equivalent of the model proposed here.

173

CHAPTER 7.   Conclusions and Future Work

WDM has been the most promising technology for long haul backbone networks. The optical technologies have matured and is slowly propagating from the core slowly towards the edge. Fiber to the home deployments have ramped up in countries like Japan and deployments are being considered currently in various European cities. With the increasing bandwidth demand in the last mile, the metro networks that bridge the access and the core are likely to go through significant developments in the near future.

WDM grooming networks have slowly evolved with time from their wavelength routing predecessors. The eventual objective of the evolution is to achieve transparent all-optical networking. This is because the speed of electronics is unlikely to be able to scale with the growing bandwidth requirements. At the same time, the individual requests from the end systems may not be large enough the fill the bandwidth of wavelength capacity. The traditional method of providing dedicated wavelength continuous path to every connection may not scale well in the future. So is the method of providing electronic grooming at the client layer due to electronic switching bottlenecks.

This brings to question whether there are alternate architectures that can accommodate dynamics in the traffic flows and yet use only off-the-shelf optical technologies. More specifically, the technology should be scalable and not be constrained by the requirement of optical switching at a packet or flow or at a burst level. The technology should allow for reconfigurability and grooming of sub-wavelength requests into big bandwidth pipes in an efficient way.

Recent research in the field of optical networks shows an increasing trend in exploring architectures that share wavelength all optically. A particular method of interest allows the lightwave circuit to be configured in the form of a simple path and shares the circuit across traffic flows that spans multiple sources and destinations. This allows for efficient packing of wavelengths without requiring active switching on a per flow basis. It allows for traffic dynamics through statistical sharing of bandwidth, albeit, at the price of increased control plane complexity.

In this dissertation, we introduce a generic framework to model path level aggregation of traffic in heterogenous WDM optical networks. We provide a virtual topology representation for a network with any path level aggregation of traffic. We consider two kinds of traffic - static and dynamic. We formulated an MILP for the static trail routing and wavelength assignment problem. We resolved the complexity class of the single-hop trail routing problem and proposed three routing and wavelength assignment heuristics of polynomial time complexity. We considered the possibility of trail sizes being limited due to physical impairments and developed algorithms that could provide results within 15 % of optimal solutions in certain network scenarios. We also studied the problem of two-hop electronic grooming and concluded that with only limited number of nodes with grooming capabilities, efficient network throughput was obtained.

We provided the formulation of an optimization problem for static survivable trail routing and wavelength assignment. We designed heuristics for connection level shared and dedicated segregated protection in light-trail based networks. We observed that with only a modest amount of spare capacity, we were able to achieve full protection for single link failures.

For the dynamic traffic scenario case, we developed an auxiliary graph based network that could model various path level aggregation schemes. We studied the impact of granularities on network design and quantified the effect of using tunable and fixed transceivers. We compared the performance of LP, SLT, DLT, and LT networks. It is seen that for router speeds less than a certain threshold, all-optical trail based networks outperform e-groomed lightpath networks. In wavelength constrained multi-hop scenarios, SLT networks perform better than LP networks. In single hop scenarios, LT networks perform better than LP networks. We observe that SLT networks offer a good trade-off between control complexity and aggregation capability.

We introduced a simple medium access control protocol for the shared medium network using optical delay lines to sense the channel and buffer the packets to prevent collisions. We modeled a small network using queuing theory and performed simulations for larger trail sizes.

There are a number of interesting research issues that arise in the context of this architecture. While a few of them have been addressed, a few are open for future work. In the case of single hop and multi-hop dynamic design, we observe that bandwidth and transceiver penalties leads to wastage of network resources. It may be interesting to design algorithms that admit a connection on a circuit only if the penalty incurred is below a certain threshold value. Having a small threshold value will degenerate the path level aggregation to LP based networks. Having a large threshold values may not lead to sufficient reduction in wastage due to bandwidth locking. The right choice for such a threshold will have to be identified.

An interesting area of further work will be in designing LT networks for multicast and groupcast communication. Such kind of communication have been achieved in the past using splitters that lead to constrained power budgets. Given a group of nodes that are involved in multicast or groupcast, it should be possible to run a linear circuit (instead of a multicast tree) such that the requirements of the multicast traffic are satisfied. Multicasting may be simple to achieve if the circuit is allowed to traverse a node more than once since there is only one source and the order of occurrence of receivers in the circuit does not matter. Groupcast, however, requires that the transmitters and receivers in the circuit be placed in certain order. For instance, if a receiver happens to be placed upstream of a transmitter, the communication between these nodes are not possible.

The aggregation schemes requires that the wavelength bandwidth be statistically shared by multiple traffic flows. The medium access protocol should be designed to ensure fairness irrespective of the position of a node on a trail. The design of such a protocol is complicated by the fact that the channel is unidirectional and though a path exists in the reverse direction in the out of band control channel, utilization may become low if feedback is required for every packet.

There are a number of issues in this field that can be researched in the future. Having identified that there is a definite advantage in employing path level aggregation, we intend to pursue this research further.

Bibliography

[1] S. Arakawa and M. Murata. Lightpath management of logical topology with incremental traffic changes for reliable IP over WDM networks. *Optical Networks Magazine*, May 2002.

[2] C. Assi, Y. Ye, A. Shami, and M. A. Ali. Integrated routing algorithms for provisioning sub-wavelength connections in IP over WDM networks. *Photonic Network Communication*, March/April 2002.

[3] D. Awduche, L. Berger, D. Gan, T. Li, V. Srinivasan, and G. Swallow. Rsvp-te: Extensions to rsvp for lsp tunnels. *RFC 3209*, December 2001.

[4] S. Balasubramanian, A.E. Kamal, and A.K. Somani. Medium access control protocols for light-trail and light bus networks. *Proceeding of 8th IFIP Working Conference on Optical Network Design and Modeling*, February 2004.

[5] S. Balasubramanian, A.E. Kamal, and A.K. Somani. Network design in IP-centric light-trail networks. *IEEE Broadnets*, October 2005.

[6] S. Balasubramanian and A.K. Somani. Traffic grooming in statistically shared optical networks. *IEEE Local Computer Networks*, September 2006.

[7] S. Balasubramanian and A.K. Somani. Design algorithms for path-level grooming of traffic in wdm metro optical networks. *OSA, Journal of Optical Networking*, August 2008.

[8] S. Balasubramanian and A.K. Somani. Path level traffic grooming in wdm metro optical networks. *IEEE Communications Magazine*, November 2008.

[9] A. Banerjee, J. Drake, J.P. Lang, B. Turner, K. Kompella, and Y. Rekhter. Generalized multiprotocol label switching: An overview of routing and management enhancements. *IEEE Communications Magazine*, January 2001.

[10] S. Banerjee and B.Mukherjee. An efficient channel feedback based adaptive protocol for scheduling variable-length messages on slotted, high-sped fiber optica lans/mans. *Proceedings of the Winter Simulation Conference*, December 1991.

[11] R.S. Barr, M.S. Kingsley, and R.A. Patterson. Grooming telecommunications network optimization models and methods. *Technical report 05-EMIS-03*, June 2005.

[12] R.S. Barr and R.A. Patterson. Grooming telecommunication networks. *Optical Networks Magazine*, May/June 2001.

[13] T. Battestilli and H. Perros. An introduction to optical burst switching. *IEEE Communications Magazine*, 41(8), August 2003.

[14] L. Berger. Generalized MPLS - signaling functional description. *RFC 3471*, January 2003.

[15] R. Berry and E. Modiano. Reducing electronic multiplexing costs in SONET/WDM rings with dynamically changing traffic. *IEEE Journal of selected areas in communications*, 18(10), October 2000.

[16] P. Bonenfant and A. Rodriguez. Framing techniques for IP over fiber. *IEEE Network*, July/August 2001.

[17] N. Bouabdallah, A. Beylot, E. Dotaro, and G. Pujolle. Resolving the fairness issues in bus-based optical access networks. *IEEE Journal of Selected Areas in Communications*, August 2005.

[18] N. Bouabdallah, E. Dotaro, L. Ciavaglia, N. Le Sauze, and G. Pujolle. Distributed aggregation in all-optical wavelength routed networks. *The 39th IEEE International Conference on Communications*, June 2004.

[19] N. Bouabdallah, H. Perros, and G. Pujolle. A cost-effective traffic aggregation scheme in all-optical networks. *IEEE GLOBECOM*, November 2005.

[20] N. Bouabdallah and G. Pujolle. Optical resource provisioning: multipoint-to-point light-paths mapping in all-optical networks. *International Journal of Network Management*, (15), March 2005.

[21] N. Bouabdallah and G. Pujolle. A practical traffic grooming scheme in all-optical networks. *IEEE Optical Fiber Conference*, March 2006.

[22] L. Calafato, M. Mellia, E. Leonardi, and F. Neri. Exploiting OTDM traffic grooming in dynamic wavelength routed networks. *Eighth working conference on Optical Network Design and Modeling*, February 2004.

[23] J. Cao. Next generation routers. *Proceedings of the IEEE*, 90(9), September 2002.

[24] A. Carena, V.D. Feo, J. M. Finochietto, R. Gaudino, F. Neri, C.Piglione, and P.Poggiolini. Ringo: An experimental WDM optical packet network for metro applications. *IEEE Journal on selected areas in communications*, 22(8), October 2004.

[25] D. Cavendish, K. Murakami, S.H.Yun, O.Matsuda, and M.Nishihara. New transport services for next generation SONET/SDH systems. *IEEE Communications Magazine*, May 2002.

[26] V. Chan and G. Weichenberg. Optical flow switching. *Invited Paper, International Workshop on Optical Burst/Packet Switching*, October 2006.

[27] B. Chen, G. Rouskas, and R. Dutta. A framework for hierarchical traffic grooming in WDM networks of general topology. *IEEE Broadnets*, October 2005.

[28] Y. Chen, C. Qaio, and X. Yu. Optical burst switching: a new area in optical networking research. *IEEE Network*, 18, May 2004.

[29] A.L. Chiu and E. H. Modiano. Traffic grooming algorithms for reducing electronic mul-tiplexing costs in WDM ring networks. *Journal of lightwave technology*, 18(1), January 2000.

[30] I. Chlamtac, A. Ganz, and G. Karmi. Lightpath communications: An approach to high bandwidth optical WANs. *IEEE Transactions on Communications*, July 1992.

[31] Y. Yeand H. Woesnerand I. Chlamtac. OTDM light trail networks. *Transparent Optical Networks*, July 2005.

[32] T. Cinkler. Traffic and lamdba grooming. *IEEE Network*, March/April 2003.

[33] J. Comellas, R. Martinez, J. Prat, V. Sales, and G. Junyent. Integrated IP/WDM routing in GMPLS based optical networks. *IEEE Network*, March/April 2003.

[34] T.H. Cormen, C.E. Leiserson, R.L. Rivest, and C.Stein. Introduction to algorithms, second edition. *Prentice Hall of India*, 2002.

[35] F. Davik, M. Yilmaz, S.Gjessing, and N.Uzun. IEEE 802.17 Resilient Packet Ring Tu-torial. *IEEE Communications Magazine*, March 2004.

[36] R. Dutta, B. Chen, and G. Rouskas. On the application of k-center algorithms in hier-archical traffic grooming. *IEEE Workshop on Traffic Grooming*, October 2005.

[37] R. Dutta and G.N.Rouskas. On optimal traffic grooming in WDM rings. *IEEE Journal on selected areas in communications*, 20(1), January 2002.

[38] R. Dutta and G. Rouskas. Traffic grooming in WDM networks: Past and Future. *IEEE Network*, November/December 2002.

[39] J. Fang, W. He, and A. K. Somani. Optimal light-trail design in WDM optical networks. *The proceedings of International Conference on Communications*, 3, June 2004.

[40] F. Farahmand, X. Huang, and J.P. Jue. Efficient online traffic grooming algorithms in WDM mesh networks with drop-and-continue node architecture. *IEEE Broadnets*, October 2004.

[41] M. T. Frederick, N. A. VanderHorn, and A. K. Somani. Light trails: a sub-wavelength solution for optical networking. *High Performance Switching and Routing (HPSR) Workshop*, pages 175–179, May 2004.

[42] A. Fredette and J. Lang. Link management protocol (LMP) for dense wavelength division multiplexing (DWDM) optical line systems. *RFC 4209*, October 2005.

[43] B. Ganguly and V. Chan. Distributed algorithms and architectures for optical flow switching in WDM networks. *Fifth IEEE Symposium on Computers and Communications*, July 2000.

[44] B. Ganguly and V. Chan. A scheduled approach to optical flow switching in the ON-RAMP optical network testbed. *Optical Fiber Conference*, March 2002.

[45] O. Gerstel, P.Lin, and G.Sasaki. Wavelength assignment in a WDM ring to minimize cost of embedded SONET rings. *IEEE Infocom*, 1998.

[46] O. Gerstel, P.Lin, and G.Sasaki. Combined WDM and SONET network design. *Proceedings of IEEE Infocom*, March 1999.

[47] O. Gerstel, R.Ramaswami, and G.Sasaki. Cost effective traffic grooming in WDM rings. *IEEE Infocom*, March 1998.

[48] N. Ghani. Metropolitan networks: Trends, technologies and evolutions. *Optical Networks Magazine*, July/August 2002.

[49] N. Ghani, J. Pan, and X. Cheng. Metropolitan optical networks. *Optical Fiber Telecommunications IV,Volume B, Academic Press*, March 2002.

[50] A. Gumaste. Optimal and heuristic assignment for light-trail assignment. *13th Symposium on Perf. Evaluation on Computers and Telecommunications*, July 2005.

[51] A. Gumaste and I. Chlamtac. Mesh implementation of light-trails: A solution to IP centric communication in the optical domain. *IEEE Proceedings of International Conference on Communication*, 2003.

182

[52] A. Gumaste and I. Chlamtac. Light-trails: an optical solution for IP transport. *Journal of Optical Networking*, April 2004.

[53] A. Gumaste, G.Kuiper, and I. Chlamtac. Optimizing light-trail assignment to WDM networks for dynamic IP-centric traffic. *The 13th IEEE Workshop on Local and Metropolitan Area Networks*, 2004.

[54] A. Gumaste, P. Palacharala, and T. Naito. Performance evaluation and demonstration of light-trails in shared wavelength optical networks (SWON). *31st European Conference on Optical Communication*, September 2005.

[55] A. Gumaste and S. Zheng. Protection and restoration scheme for light-trail WDM ring networks. *9th conference on Optical Network Design and Modelling*, February 2005.

[56] U. Gupta, D. Lee, and J. Leung. Efficient algorithms for interval graphs and circular-arc graphs. *Networks*, 12, 1982.

[57] A.M. Hamad and A.E. Kamal. Optimal power-aware design of all-optical multicasting in wavelength routed networks. *IEEE International Conference on Communications*, 3, June 2004.

[58] N.F. Huang, G.-H. Liaw, and C.-P. Wang. A novel all-optical transport network with time-shared wavelength channels. *IEEE Journal of Selected Areas in Communications*, 18(10), October 2000.

[59] S. Huang and R. Dutta. Research problems in dynamic traffic grooming in optical networks. *Workshop on Traffic Grooming, IEEE Broadnets*, October 2005.

[60] X. Huang, F.Farahmand, and J.P.Jue. An algorithm for traffic grooming in WDM mesh networks with dynamically changing light-trees. *IEEE Globecom*, November/December 2004.

[61] R. Jain. IP over WDM networks. *Class Lectures on Recent Advances in Networking*, 1999.

[62] N.K. Jaiswal. Priority queues. *Academic Press, New York*, 1968.

[63] B. Jamoussi, L. Andersson, R. Callon, R. Dantu, L. Wu, P. Doolan, T. Worster, N. Feldman, A. Fredette, M. Girish, E. Gray, J. Heinanen, T. Kilty, and A. Malis. Constraint-based lsp setup using ldp. *RFC 3212*, January 2002.

[64] F. Jia, B. Mukherjee, and J. Iness. Scheduling variable length messages in a single-hop multichannel local lightwave network. *IEEE/ACM Transactions on Networking*, 3(4), August 1995.

[65] A.E. Kamal. Algorithms for multicast traffic grooming in WDM mesh networks. *IEEE Communications, Optical Communications Series*, 44(11), November 2006.

[66] A.E. Kamal and R. Ul-Mustafa. Multicast traffic grooming in WDM networks. *Proceedings of Opticomm 2003*, October 2003.

[67] P. Kamath, J. D. Touch, and J. A. Bannister. The need for media access control in optical CDMA networks. *IEEE Infocom*, March 2004.

[68] L. Kleinrock. Queueing systems. Volume I,II. *John Wiley and Sons, New York*, 1975.

[69] D. Kliazovich, F. Granelli, H.Woesner, and I. Chlamtac. Bidirectional light-trails for synchronous communications in WDM networks. *IEEE Globecom*, November 2005.

[70] M. Kodialam and T.V. Lakshman. Integrated dynamic IP and wavelength routing in IP over WDM networks. *IEEE Infocom*, April 2001.

[71] K. Kompella and Y. Rekhter. Intermediate system to intermediate system (IS-IS) extensions in support of generalized multi-protocol label switching (gmpls). *RFC 4205*, October 2005.

[72] K. Kompella and Y. Rekhter. OSPF extensions in support of generalized multi-protocol label switching (GMPLS). *RFC 4203*, October 2005.

[73] V.R. Konda and T.Y. Chow. Algorithm for traffic grooming in optical networks to minimize the number of transceivers. *IEEE Workshop on High Performance Switching and Routing*, May 2001.

[74] S. Koo, G. Sahin, and S. Subramanian. Cost efficient LSP protection in IP/MPLS-Over-WDM overlay networks. *IEEE International Conference on Communications*, May 2003.

[75] S. Koo, G. Sahin, and S. Subramanian. Dynamic LSP provisioning in overlay, augmented and peer architectures for IP/MPLS over WDM networks. *IEEE Infocom*, March 2004.

[76] J. Li, G. Mohan, E. Cheng, and K.C. Chua. Dynamic routing with inaccurate link state information in integrated IP over WDM networks. *Elsevier Computer Networks*, December 2004.

[77] M. Listani, V. Eramo, and R. Sabella. Architectural and technological issues for future optical internet networks. *IEEE Communications Magazine*, September 2000.

[78] M. Mellia, E. Leonardi, M. Feletig, and F. Neri R. Gaudino. Exploiting OTDM technology in WDM networks. *Proceedings of Infocom*, June 2002.

[79] E. Modiano and P.Lin. Traffic grooming in WDM networks. *IEEE Communications Magazine*, July 2001.

[80] G. Mohan and C.Murthy. Lightpath restoration in WDM optical networks. *IEEE Network*, November/December 2000.

[81] B. Mukherjee and A.E. Kamal. Scheduling variable length messages on slotted, high speed fiber optic LANs/MANs using the continuation-bit approach. *IEEE Infocom*, April 1991.

[82] B. Mukherjee and J.S. Meditch. The pi-persistent protocol for unidirectional broadcast bus networks. *IEEE Transactions on Communications*, 36(12), December 1988.

[83] L. Noirie, M. Vigoureus, and E. Dotaro. Impact of intermediate traffic grouping on the dimensioning of multi-granularity optical networks. *Proceedings Optical Fiber Communications*, March 2001.

[84] C.S. Ou, K. Zhu, H.Zhang, L.H.Sahasrabuddhe, and B.Mukherjee. Traffic grooming for survivable WDM networks: shared protection. *IEEE Journal of Selected Area in Communications*, 21(9), November 2003.

[85] C.S. Ou, K. Zhu, J.Zhang, H.Zhu, B.Mukherjee, H.Zang, and L.H.Sahasrabuddhe. Traffic grooming for survivable WDM networks: dedicated protection [invited]. *Journal of Optical Networking*, 3(1), January 2004.

[86] P. Palacharla, A. Gumaste, E. Biru, and T. Naito. Implementation of burstponder card for ethernet grooming in light-trail WDM networks. *41st IEEE International Conference on Communications*, June 2006.

[87] P. Petracca, M. Mellia, E. Leonardi, and F. Neri. Design of WDM network exploiting OTDM and light splitters. *Second International Workshop on QoS in multiservice IP Networks*, February 2003.

[88] C. Qiao and M. Yoo. Choices, features and issues in optical burst switching. *SPIE Optical Networks Magazine*, April 2000.

[89] B. Rajagopalan, D. Pendarakis, D. Saha, R.S. Ramamoorthy, and K. Bala. IP over optical networks: Architectural aspects. *IEEE Communications Magazine*, September 2000.

[90] R. Ramaswami. Optical networking technologies: what worked and what didn't. *IEEE Communications Magazine*, 44(9), September 2006.

[91] R. Ramaswami and G.H.Sasaki. Multiwavelength optical networks with limited wavelength conversion. *IEEE/ACM Transaction on Networks*, 6(6), December 1998.

[92] R. Ramaswami and K.N. Sivarajan. Optical networks: A practical perspective. *Second edition, Morgan Kaufmann publishers*, 2002.

[93] E. Rosen, A. Viswanathan, and R. Callon. Multiprotocol label switching architecture. *RFC 3031*, January 2001.

[94] L. Ruan and F. Tang. Survivable IP network realization in IP-over-WDM networks under overlay model. *Computer Communications*, 29(10):1772–1779, June 2006.

[95] L. H. Sahasrabuddhe and B. Mukherjee. Light-trees: Optical multicasting for improved performance in wavelength-routed networks. *IEEE Communication Magazine*, February 1999.

[96] J. Salehi and C. Brackett. Code division multiple-access techniques in optical fiber networks - part 1: Fundamental principles. *IEEE Transactions on Communications*, August 1989.

[97] E. Salvadori, R.L. Cigno, and Z. Zsoka. Dynamic grooming in IP over WDM networks: A study with realistic traffic based on GANCLES simulation package. *IFIP Optical Network Design and Modeling*, February 2005.

[98] G. H. Sasaki and C.F. Su. The interface between IP and WDM and its effect on the cost of survivability. *IEEE Communications Magazine*, pages 74–79, January 2003.

[99] M. Scholten, Z.Zhu, E.H.Valencia, and J. Hawkins. Data transport applications using GFP. *IEEE Communications Magazine*, May 2002.

[100] S. Sheeshia and Chunming Qiao. Burst grooming in Optical-Burst-Switched networks. *IEEE Broadnets Traffic Grooming Workshop*, October 2004.

[101] J.M. Simmons, E.L.Goldstein, and A.A.M.Shah. Quantifying the benefit of wavelength add-drop in WDM rings with distance-independent and dependent traffic. *Journal of Lightwave Technology*, 17(1), January 1999.

[102] M. Sivakumar and S. Subramaniam. Blocking performance of time switching in TDM wavelength routing networks. *Optical Switching and Networking*, February 2005.

[103] N. Srinivas and C. S. R. Murthy. Design and dimensioning of a WDM mesh network to groom dynamically varying traffic. *Photonic Network Communications*, 7(2), March 2004.

[104] R. Srinivasan. MICRON - a framework for connection establishment in optical networks. *Opticomm*, October 2003.

[105] R. Srinivasan and A.K. Somani. A generalized framework for analyzing time-space switched optical networks. *IEEE Journal of Selected Areas in Communications*, January 2002.

[106] S. Subramaniam, E. J. Harder, and H.-A. Choi. Scheduling multi-rate sessions in TDM wavelength-routing networks. *Proceedings of GLOBECOM*, December 1999.

[107] H. Takagi. Queueing Analysis. Volume I. *New York, Elsevier Science Publishers*, 1991.

[108] Y. Takushima and K. Kikuchi. Photonic switching using spread spectrum technique. *IEEE Electronics Letters*, March 1994.

[109] A. N. Tam, P.J. Lin, and E. Modiano. Efficient routing and wavelength assignment for reconfigurable WDM networks. *IEEE Journal of Selected Areas in Communications*, January 2002.

[110] V. Tamilraj and S. Subramaniam. An analytical blocking model for dual-rate sessions in multichannel optical networks. *Proceedings of GLOBECOM*, November 2001.

[111] S. Thiagarajan and A. K. Somani. A capacity correlation model for WDM networks with constrained grooming capabilities. *IEEE International Conference on Communications*, December 2001.

[112] S. Thiagarajan and A. K. Somani. Capacity fairness of WDM networks with grooming capabilities. *Optical Networks Magazine*, May/June 2001.

[113] TR-NWT-000253. Synchronous optical network (SONET) transport systems: Common generic criteria. *Bellcore*, January 1999.

[114] J. Turner. Terabit burst switching. *Journal of High Speed Networks*, 8(1), March 1999.

[115] E.H. Valencia, M.Scholten, and Z. Zhu. The generic framing procedure: An overview. *IEEE Communications Magazine*, May 2002.

[116] N. Vanderhorn, S. Balasubramanian, M. Mina, B. R. Weber, and A. K. Somani. Light-trail test bed for metro optical networks. *IEEE Communications Magazine*, May 2003.

[117] N. A. VanderHorn, M.Mina, and A.K. Somani. Light-trails: A passive optical networking solution for wavelength sharing in the metro. *Workshop on High Capacity Optical Networks and Enabling Technologies*, December 2004.

[118] S. Verma, H. Chaskar, and R. Ravikanth. Optical burst switching: A viable solution for terabit IP backbone. *IEEE Network*, 14(6), November/December 2000.

[119] B. Wang, T. Li, X. Luo, and Y. Fan. Multicast service provisioning under a scheduled traffic model in WDM optical networks. *IEEE Workshop on traffic grooming in Broadnets*, October 2004.

[120] J. Wang, V.R.Vemuri, W.Cho, and B.Mukherjee. Improved approaches for cost-effective traffic grooming in WDM ring networks:nonuniform traffic and bidirectional ring. *Proceedings of International Conference on Communications*, 3, June 2000.

[121] J. Wang, W.Cho, V.R.Vemuri, and B.Mukherjee. Improved approaches for cost-effective traffic grooming in WDM ring networks:ILP formulations and single-hop and multi-hop connections. *Journal of Lightwave Technology*, 19(11), November 2001.

[122] B. Waxman. Routing of multipoint connections. *IEEE Journal of Selected Areas in Communications*, December 1988.

[123] B. Wen and K. Sivalingam. Routing, wavelength and time slot assignment in time division multiplexed wavelength routed networks. *Proceedings of Infocom*, June 2002.

[124] I. Widjaja, I.Saniee, R.Giles, and D.Mitra. Light core and intelligent edge for a flexible, thin-layered and cost-effective optical transport network. *IEEE Communications Magazine*, 41(5), May 2003.

[125] G. Wilfang and P.Winkler. Ring routing and wavelength translation. *Proc. of 9th Annual ACM-SIAM Sump. Discrete Algorithms*, 1998.

[126] R.W. Wolff. Stochastic modeling and the theory of queues. *Prentice Hall, Englewood Cliffs, NJ*, 1988.

[127] C. Xin and C. Qiao. A comparative study of OBS and OFS. *Optical Fiber Conference*, March 2001.

[128] C. Xin and C. Qiao. Performance analysis of multi-hop traffic grooming in mesh WDM optical networks. *IEEE International Conference on Computer Communication and Networks*, October 2003.

[129] Y. Xin, G. N. Rouskas, and H. G. Perros. On the physical and logical topology design of large-scale optical networks. *Journal of Lightwave Technology*, 21(4), April 2003.

[130] L. Xu, H. Perros, and G. Rouskas. A survey of optical packet switching and optical burst switching. *IEEE Communication Magazine*, 39, 1 2001.

[131] M. Pioro Y. Brehon, D. Kofman and M. Diallo. Optimal virtual topology design using bus-label switched paths. *IEEE Journal of Selected Areas in Communications*, June 2007.

[132] G. N. Rouskas Y. Xin. A study of path protection in large-scale optical networks. *Photonic Network Communications*, 7(4), May 2004.

[133] M. Yao, M. Li, and B. Ramamurthy. Performance analysis of sparse traffic grooming in WDM networks. *IEEE International Conference on Communications*, May 2005.

[134] S. Yao, B. Mukherjee, and S. Dixit. Advances in photonic packet switching: an overview. *IEEE Communication Magazine*, 38, February 2000.

[135] T. Ye, Y. Su, L. Leng, Q. Zeng, and Y. Jin. SLEA: A novel scheme for routing in overlay IP/WDM networks. *IEEE Journal of Lightwave Technology*, 23, October 2005.

[136] T. Ye, Q. Zeng, Y. Su, L. Leng, W. Wei, Z. Zang, W. Guo, and Y. Jin. On-line integrated routing in dynamic multifiber IP/WDM networks. *IEEE Journal of Selected Areas in Communications*, 22(9), November 2004.

[137] Y. Ye, C. Assi, S. Dixit, and M.A. Ali. A simple dynamic integrated provisioning/protection scheme in IP over WDM networks. *IEEE Communications Magazine*, November 2001.

[138] Y. Ye, H.Woesner, R.Grasso, T.Chen, and I.Chlamtac. Traffic grooming in light trail networks. *IEEE Globecom*, November 2005.

[139] M. Yoo and C.Qiao. Optical burst switching (OBS) - a new paradigm for the optical internet. *Journal of High Speed Networks*, 8(1), 1999.

[140] W. Zhang, G.Xue, J.Tang, and K.Thulasiraman. Dynamic light trail routing and protection issues in wdm optical networks. *IEEE Globecom*, November 2005.

[141] X. Zhang and C.Qiao. An effective and comprehensive approach for traffic grooming and wavelength assignment in SONET/WDM rings. *IEEE/ACM Transactions on Networking*, 8(5), October 2000.

[142] D. Zhemin, M. Hamdi, and JYB Lee. Integrated routing and grooming in GMPLS-based optical networks. *IEEE International Conference on Communications*, June 2004.

[143] Q. Zheng and G. Mohan. An efficient dynamic protection scheme in integrated IP/WDM networks. *IEEE International Conference on Communications*, May 2003.

[144] H. Zhu, H. Zang, K. Zhu, and B. Mukherjee. A novel generic graph model for traffic grooming in heterogenous WDM mesh networks. *IEEE/ACM Transactions on Networking*, April 2003.

[145] H. Zhu, H. Zang, K. Zhu, and B. Mukherjee. Dynamic traffic grooming in WDM mesh networks using a novel graph model. *IEEE Globecom*, November/December 2004.

[146] K. Zhu, H.Zang, and B.Mukherjee. A comprehensive study of next-generation optical grooming switches. *IEEE Journal of selected areas in communications*, 27(7), September 2002.

[147] K. Zhu, H.Zang, and B. Mukherjee. Design of WDM networks with sparse grooming capability. *IEEE Globecomm*, November 2002.

[148] K. Zhu and B. Mukherjee. On-line approaches for provisioning connections of different bandwidth granularities in WDM mesh networks. *Optical Fiber Communication Conference*, March 2002.

[149] K. Zhu and B. Mukherjee. Traffic grooming in an optical WDM mesh network. *IEEE Journal of Selected Areas in Communiations*, January 2002.

[150] K. Zhu, H. Zhu, and B.Mukherjee. Traffic engineering in multigranularity heterogeneous optical WDM mesh networks through dynamic traffic grooming. *IEEE Network*, 17(2), March/April 2003.